Heat Transfer 4

Mathematical and Mechanical Engineering Set

coordinated by
Abdelkhalak El Hami

Volume 12

Heat Transfer 4

Convection, Two-Phase Flows and Special Problems

Michel Ledoux
Abdelkhalak El Hami

iSTE

WILEY

First published 2022 in Great Britain and the United States by ISTE Ltd and John Wiley & Sons, Inc.

ISTE Ltd
27-37 St George's Road
London SW19 4EU
UK

www.iste.co.uk

John Wiley & Sons, Inc.
111 River Street
Hoboken, NJ 07030
USA

www.wiley.com

Library of Congress Control Number: 2022941257

British Library Cataloguing-in-Publication Data
A CIP record for this book is available from the British Library
ISBN 978-1-78630-879-5

Contents

Preface

Thermal science is to thermodynamics as decree is to law. It answers the following question – which all good leaders must (or should) ask themselves whenever they have an "idea": "How would this work in practice?". In a way, thermal science "implements" thermodynamics, of which it is a branch. A thermodynamics specialist is a kind of energy economist. Applying the first principle, they create a "grocery store". With the second principle, they talk about the quality of their products. I add or remove heat from a source or work from a system. And the temperature, among other things, defines the quality of the energy for me.

But *by what means do I take or do I give*? Even calculations of elementary reversible transformations do not tell us by what process heat passes from a source to a system.

Thermal science specifies how, but "evacuates" work. If in a given problem related to, for example, a convector where an electrical energy (therefore in the "work" category) appears, it is immediately dissipated into heat by the Joule effect.

Three heat transfer modes can be identified: conduction and radiation – which can be seen separately, although they are often paired up – and convection, which is by nature an interaction of fluid mechanics and conduction.

This volume, like *Heat Transfer 3*, studies thermal science in three parts, which are distributed into four volumes. This separation seemed necessary to us in order to keep a reasonable size to the volumes of the collection.

Heat Transfer 4 focuses on specific convection transfers. The opening on two-phase transfer is two-fold. Notions on bubble liquid flows and dispersed flows (sprays) are presented. A more recent and far less known domain, namely that of phase change transfer, is presented. Specific attention is paid to numerical application, giving the reader the opportunity to handle orders of magnitude, which is a significant aspect in physics.

This work is intended to reach a wide audience, from technicians to engineers, to researchers in many disciplines, whether physicists or not, who have a one-off transfer problem to resolve in a laboratory context. With this in mind, the theoretical developments in the text itself are as direct as possible. Specialist readers, or those who are simply curious about further theoretical developments (general equations, particular problems, mathematical tools, etc.) may refer to the Appendices.

This fourth volume focusing on the “classical” (analytical) approaches to convection may be of higher interest to readers looking for “simple” prediction modes.

It is organized in four chapters.

Chapter 1 is dedicated to two-phase transfers. This very complex domain, whose technological significance is obvious, is not presented in all the textbooks. Some aspects related to the physics of fluids under evaporation are also presented. This chapter also provides an introduction to the subject of sprays, which is generally insufficiently addressed in classical literature.

Chapter 2 explores the domain of mass convection. This is not a minor point in a book dedicated to transfer, as it is important in the chemical engineering-type of approaches. This chapter highlights the analogy between heat and mass transfer equations.

Chapter 3 is an introduction to the domain of phase change materials. It explores storage, which is often associated with and is downstream of convective phenomena.

Chapter 4 provides an introductory presentation of thermomechanical interaction problems. A heat engineer cannot do without this subject as it is of practical importance, and is still insufficiently present in the introductory literature.

The detailed calculation of the establishment of fundamental equations is quite burdensome. **Appendices 1**, **2** and **3** summarize several physical data, and finally, **Appendix 4** recalls the values of function $erf(x)$ and of its associates, which are sometimes useful for certain calculations.

August 2022

Introduction

I.1. Preamble

Thermal energy was probably first perceived (if not identified) by humanity, through the Sun. The themes of night and day are found at the center of most ancient myths. Humanity's greatest fear was probably that the Sun would not return again in the morning. Fire became controlled in approximately 400,000 BP. Thermal transfer was therefore a companion of *Homo ergaster*, long before *Homo sapiens sapiens*.

However, it took a few hundred thousand years before so-called "modern" science was born. Newtonian mechanics dates from three centuries ago. Paradoxically, another century and a half passed by before energy was corrections perceived by scientists, in terms of the new field of thermodynamics. Furthermore, a systematic study of heat transfer mechanisms was carried out at the end of the 19th century, and even later for the study of limit layers, the basis of convection.

Heating, lighting and operating the steam engines of the 19th century were all very prosaic concerns. Yet this is where revolutions in the history of physics began: the explosion of statistical thermodynamics driven by Boltzmann's genius, and quantum mechanics erupted with Planck, again with Boltzmann's involvement.

Advances in radiation science, particularly in sensor technology, has enabled us to push back our "vision" of the universe by a considerable number of light years. To these advances we owe, in particular, the renewed interest in general relativity that quantum mechanics had slightly eclipsed, through demonstration of *black holes*, the physics of which may still hold further surprises for us.

Closer to home, fundamental thermal science, where it is conduction, convection or radiation, contributes to the improvement of our daily lives. This is particularly true in the field of housing where it contributes, under pressure from environmental questions, to the evolution of new concepts such as the active house.

The physics that we describe in this way, and to which we will perhaps introduce some readers, is therefore related both to the pinnacles of knowledge and the banality of our daily lives. Modestly, we will place our ambition in this latter area.

There are numerous heat transfer textbooks in different formats: "handbooks" attempting to be exhaustive are an irreplaceable collection of correlations. High-level courses, at universities or engineering schools, are also quite exhaustive, but they remain demanding for the listener or the reader. Specialist, more empirical thematic manuals are still focused on specialists in spite of all this.

So why do we need another book?

The authors have taught at university level and in prestigious French engineering schools, and have been involved in the training of engineers on block-release courses. This last method of teaching, which has been gaining popularity in recent years, particularly in Europe, incorporates a distinctive feature from an educational point of view. Its practice has, in part, inspired this book.

The aim is to help learners who have not had high-level mathematical training in their first years following the French Baccalaureate (therefore accessible to apprentices), and pupils with more traditional profiles. At the same time, we would like to show this broad audience the very new possibilities in the field of digital processing of complex problems.

When a miner wants to detach a block of coal or precious mineral from a wall, they pick up a pneumatic drill. If we want to construct a tunnel, we must use dynamite. The same is true for physicists.

Whether they are researchers, engineers or simply teachers, scientists have two tools in their hands: a *calculator* and a *computer* (with very variable power). Since both authors are teacher researchers, they know they owe everything to the invention of the computer. From the point of view of teaching, however, each one of the two authors has remained specialist, one holding out for the calculator and "back-of-the-napkin" calculations, and the other one using digital calculations.

The revolution that digital tools has generated in the world of "science" and "technical" fields, aside from the context of our daily lives, no longer needs to be proved. We are a "has been" nowadays if we do not talk about Industry 4.0. The "digital divide" is bigger than the social divide, unless it is part of it…

Indeed, the memory of this revolution is now fading. Have students today ever had a “slide rule” in their hands? Do they even know what it is? Yet, all the physicists behind the laws of thermal science had only this tool in hand, giving three significant figures (four with good visibility and tenacity), leaving the user to find a power of ten of the result. It goes without saying that a simple calculation of a reversible adiabatic expansion became an ordeal, which played a part in degrading the already negative image of thermodynamics held by the average student.

This reminder will seem useless to some; slide rules are at best sleeping in drawers. But there is a moral to this story: no matter what type of keyboard we type on, a calculator or a computer, our head must have control over our fingers. This book has been written on the basis of this moral.

A good physicist must have a perfect understanding of the idea of an “order of magnitude”. For this, the tool is a calculator. We always do a rough sizing of a project before moving on to detailed modeling and numerical calculations.

The two authors belong to the world of engineering sciences, meaning most of their PhD students have entered the private sector. One of them, having moved into the aerospace sector, came back to see us very surprised by the recurrence of “back-of-the-napkin” calculations in his day-to-day work.

Fundamental or “basic physics” concepts are taken from a type of manual that is resolutely different from those dedicated to the numerical approach. In this case, the authors allow themselves to believe that it is no bad thing to collect them all together in a single book, for once. This is a significant difference that will surprise some and, without doubt, be criticized by others. Nevertheless, when reading this book, an “average” student will be initiated to a field that teaching models generally promised “for later on” (or never if they never go beyond a certain level of education). It is also true that fully immersed in equations and complex calculations, specialist readers will be able to “be refreshed” when faced with the short exercises, which can sometimes surprise and encourage them – why not – to go back to their roots (assuming they had indeed been there).

Another significant difference is that this book is directed at a large scientific audience, which covers possibly the entire field: researchers, PhD students or those who have obtained Confirmation and are just starting out in the field, technicians, students or professionals, engineers. This last type of scientist is perhaps the main target of this book.

Dividing the study of thermal science into three volumes is the result of logic. Presenting this work in four volumes is somewhat arbitrary; in our opinion, however, this split was necessary in order to keep the volumes in the collection a reasonable size.

The first volume, entitled **Heat Transfer 1**, is dedicated to "classic" approaches (analytical treatment) to conduction, which will be of greater interest to readers who are looking for "simple" prediction methods.

The second volume, entitled **Heat Transfer 2**, is dedicated to "classic" approaches (analytical treatment) of radiation, and assembles digital approaches of these various transfer modes. It is aimed at engineers or researchers who want to resolve more complex problems.

The third and fourth volumes, entitled **Heat Transfer 3** and **Heat Transfer 4** are focused on convection transfers. **Heat Transfer 3** deals with the fundamentals, integrating various modes of approach, both empirical and theoretical (boundary layer), and gives an introduction to the theory of exchangers. As we have already pointed out, all of these transport operations are rarely pure and lead to problems that involve three inter-connecting transfer modes, conduction, convection and radiation.

Heat Transfer 4 aims to broaden the reader's horizon to more complex transfer modes, such as two-phase transfers. It looks at mass transfer, often in analogy with heat transfer, and explores less-known fields, such as phase change materials. It also introduces the electro-thermo-mechanical modeling of systems.

So, what is this book for?

Above all, it **contains problems to be worked on**, of which most are accessible to all, from the level of an apprentice technician upwards, either one or two years after the Baccalaureate. This book was written in France, where scientific teaching is structured around universities, engineering Grandes Écoles, engineering training through apprenticeships and two types of technician training sections at high schools or universities. In countries with simpler models, readers should also find it useful.

It seems necessary to surround these problems with strong reminders of past learning, so that the reader does not need to permanently refer back to their manuals. We see two advantages in this: a presentation of the scientific material focusing on the problems, and a second chance for readers to integrate notions that perhaps had not been well understood in the initial teaching.

Lastly, upon rereading, the authors also recommend this book as an introduction to the taught disciplines.

I.2. Interlude

Before our readers immerse themselves in a text that, despite our best efforts, remains intellectually demanding, we propose a short text that is a little lighter.

This does not mean that it is not significant in terms of understanding the physics behind all the calculations proposed in this book.

Let us imagine, in a "B movie" context, a somber hostel in the gray fog of a port in the middle of nowhere. Sailors from a faraway marina come and drink away their troubles. And as always, the drink helping them along, they turn to fighting.

Let us entrust Ludwig Boltzmann to the direct the film. Our B movie heroes are getting agitated, delivering blows to one another. Each one of them has moderate kinetic energy, distributed heterogeneously among them in the room. For some reason, they get involved in a general brawl. Their average kinetic energy becomes much greater.

In everyday language, we would say that ***things are hotting up.***

This would bring us right into line with a fundamental concept of Boltzmann, who was the first to hypothesize that ***heat is made up of molecular agitation****. The temperature in a gas is proportional to the average quadratic energy of the molecules that make it up.*

$$E_C = \frac{1}{2} kT$$

Using this model, we will return to the physical basis for all transport phenomena.

On the way, we rarely escape from the explosion of a door or a window, giving in under the repeated beatings of the brawlers.

We have just modeled the pressure, due to the transfer of the quantity of movement on the surface, by the impact of molecules.

Let us now imagine that the altercation is initially located in the corner of the room: a smaller group starts fighting between themselves.

From kicks to punches, after multiple impacts within the group and its immediate neighbors, the agitation will spread: *we have just seen the mechanism of heat propagation by transfer of impacts.*

Let us place an imaginary separation (geometrically but immaterially defined) at the center of the room. Let us count the sailors that cross through it within a unit of time.

This wall is now crossed by kinetic energy: we have defined a flow of heat.

Let us put a metal ring with a surface area of $S = 1\,m^2$ in the room. On both sides of this ring, the blows exchanged constitute a transfer of kinetic energy – we have just defined **the heat flow density**.

And we have just understood the nature of the propagation of thermal flows by impacts.

Let us suppose that the great majority of the brawlers come from a ship with a white uniform. Let us suppose that another boat in the port has uniforms that are red. The red ones are initially all united. We will then quickly see that the red mariners, as they receive and return blows, spread out across the room. *We have just shown the mechanism of **diffusion of matter**, of a component within a mixture.*

We will have a better qualitative understanding that the fundamental law of conduction (Fourier Law) is formally identical to the law for the diffusion of mass (Fick Law).

Let us put our agitated sailors in the compartments of a flatcar train, where they continue to fight. And let us start the train moving. The kinetic energy that they contain is transported from one point to another.

*We have just invented **thermal convection**.*

We can go further.

Let us imagine a series of flatcar trains on a set of parallel tracks. The train furthest to the side is fixed to a platform. All of these trains are full of sailors. Let us suppose that our train follows the outside, parallel rail tracks. No brakes will prevent these trains from moving. Only the last train, at the platform, is stuck.

For a reason we do not need to analyze (cinema allows all kinds of fantasy), "clusters" of fighting sailors jump from one wagon to the next. These "clusters" contain a component of speed that is parallel to the train, which will communicate information about the quantity of movement to the adjacent train. These trains will then start to move, more quickly the closer they are to the outside train. And the same occurs up to the train at the platform. This train will not move, but a force will be applied to its brakes.

We have just discovered the mechanism of dynamic viscosity. At the same time, the parallel trains in relative movement give us a picture of the *notion of boundary layers.*

At the same time, these agitated clusters carry their disordered kinetic energy, "thermal" agitation. We have just seen the mechanism of the *thermal boundary layer.*

Finally, let us include a few red mariners in the crowd of white. They will be carried with the clusters, and we have just invented *the limit layer of diffusion of a species.*

We are in a fantasy, and let us benefit from it as far as we can. To finish, let us suppose that this is carnival day; each sailor has a belt equipped with bells.

All the individuals have a different speed, and the impacts are random, all the bells start to jingle, each with a different frequency. The distribution of frequencies will depend on the statistical distribution of speeds (Boltzmann statistics), and the intensity of noise produced will depend on the total agitation energy of the sailors.

We have just understood the ***basic mechanism of radiation****. We have just realized why the theory of radiation needed to use the concepts of statistics from the work of Boltzmann – a brilliant pupil of Planck – to produce the emissions spectrum of a black body, for example.*

NOTE.– *The model is certainly simplistic. The emission comes from quantum transitions in the gas atoms.*

Here, we have already deviated from the pure substance of the book, but we could go even further.

Let us suppose that our agitated sailors are in a room with one mobile wall (a nightmare scenario frequently seen on the silver screen).

The incessant impacts of the fights on this wall create a force that pushes it. This force, reduced to a surface unit, explains the notion of pressure.

By pushing against this wall, our crowd applies work that is greater than the resistance.

Here we see an equivalence spring up between work and heat that, at a fundamental level, are simply two mechanical energies: one ordered and the other disordered. **The first principle of thermodynamics is illustrated by this.**

We can see that the incidence of an average blow on the wall is rarely normal.

Therefore, an average fighter will have a trajectory that will be reflected off the wall. And only the normal component of its speed will be able to push (or transfer work to) the wall.

Thus, we see that it will be impossible for the crowd (taken to mean a gas) to give all its energy to a mobile wall.

The fundamental mechanism that leads to the **second principle of thermodynamics** *has just been demonstrated.*

These "light-hearted" images, which will perhaps not please everyone, were an oral support for the presentation of different transport phenomena by one of the authors. We hope that the reader, once they have studied this book, will want to return to this text. They will then have understood, we hope, the images that lead to the development of thermodynamics.

And if this text has a moral, it would be: *Writing down thermodynamics, just like thermal science, is based on continuous equations. The fundamentals of physics that determine these phenomena arise from the field of the discontinuous: discontinuity of matter, divided into particles; discontinuity of light, divided into photons.*

List of Notations

Some notations that are systematically used in this book are defined below.

L	Length
l	Width
H, h	Height
s, S	Surface
e	Thickness
x	Abscissa
y	Ordinate
z	Dimension (third dimension)
T	Temperature
c	Mass fraction
ρ	Mass density
μ	Dynamic viscosity
ν	Kinematic viscosity
c	Calorific mass capacity

λ	Thermal conductivity
a	Thermal diffusivity
D_i	Diffusion coefficient (species i)
Pr	Prandtl number
Sc	Schmidt number
h	Convection coefficient
R_L, R_D, R_e, R_x, etc.	Reynolds number
Nu_L, Nu_x, Nu_D, Nu_e, etc.	Nusselt number
St_x, St_L	Stanton number
Sh_L, SH_x, Sh_D, Sh_e, etc.	Sherwood number
St_{Diffx}, St_{DiffL}	Diffusive Stanton number
Pe	Péclet number
Φ	Thermal flow
φ, φ_W	Thermal flow density
$\dot{m}$	Mass flow
J	Mass flow density
R_{th}	Thermal resistance
k	Thermal resistance (exchangers)
k	Chemical kinetics

1

Concepts of Gas–Liquid Flows

1.1. Introduction and basic concepts

Two-phase convection is a specific domain of convective transfers. It is mainly of interest for two industrial fields, the chemical industry and especially the petrochemical industry, where distillation is the basic process. It is also of interest for the nuclear industry, which for several decades has catalyzed the research in this field. Nuclear research centers (such as CENG in France) and global electricity producers (EDF) are leaders in this field. For this industry, transfer efficiency is the objective.

The interest of the two-phase transfer will be briefly explained. Increasing the temperature of one kilogram of water by $\Delta T = 100\,C$ requires an amount of heat $\Delta Q_1 = 4180 * 100 = 4.18.10^6\ J$. Evaporating the same mass requires $\Delta Q_2 = m\ L = 2.09.10^6\ J$, which adds to the heat transfer when the heated water is evaporated.

This theoretical and practical interest is compensated by an increased complexity of the phenomena and their prediction.

The authors feel that it advisable to provide a overview, if not a culture, on this specific field, which is important in terms of the technical and economic stakes of the subject. Even less so than in other contexts, this overview does not pretend to be exhaustive, or to serve as training for specialists.

For a color version of all the figures in this chapter, see: www.iste.co.uk/ledoux/heat4.zip.

Any reader interested in further information should dive into the rich but complex literature. The reference section to this book includes a deliberately limited number of references, which provide a solid basis and a convenient starting point for conducting research that may prove to be full of surprises.

Since the reader may lack the skills required for the calculation of a nuclear power plant, this part of the book focuses on the presentation of the basic concepts.

A distinction must be made between liquid and vapor flows, in which gas is generated in the liquid phase, and sprays, in which the liquid phase is generated externally and carried by a gas flow.

1.2. Gas–liquid flows

The following presentation focuses on forced convection in a vertical tube. The research and design of installations involve the consideration of various questions:

a) From the point of view of physics and thermodynamics

– The experiments for the characterization of two-phase flows generally proceed by the simultaneous introduction of a piping by two flow rates of liquid and gas determined by the experimenter.

– In the heat transfer, vapor results from heat input. The "quality" is determined by the transfer phenomenon, of which it is an integral part.

The fairly complex "boiling" processes must then be taken into account first. Other interacting phenomena are those related to the gas–liquid equilibrium and capillarity. They are dealt with in the section introducing the concept of nucleate, saturated or unsaturated flow.

b) From the point of view of fluid mechanics

– The coexistence of two phases, liquid and vapor (often vapor of liquid), requires a reconsideration of the flow structure. Various structures notably appear as a function of the flow "quality" (this term will be explained in what follows). An a priori distinction must be made between ascending vertical flows, descending vertical flows and horizontal flows. It is also worth mentioning, without going into detail, the case of annular flows, which occur in certain heat exchangers.

– Every flow of real fluid involves a loss of energy by friction, which is referred to as head loss in fluid mechanics. Here again, the calculation of head loss of a two-phase flow through the pipe (as well as in annular space) is needed.

c) From the point of view of heat transfer between a fluid and the wall of a piping system

– A general theme of the problem is presented by relying on the "pool boiling" phenomenon; see section 1.2.1.4.

– Then the calculation of diffusive convective heat transfer is approached. The presentation is deliberately focused on the seemingly most common case of a vertical tube with constant heat load.

Beyond acquiring "basic knowledge" on the subject, the reader is expected to be motivated to follow a literature research on an exciting but complex subject.

1.2.1. *The physics of two-phase flows*

1.2.1.1. *Review of thermodynamics: liquid–vapor equilibrium*

The following is a review of a thermodynamics lecture dedicated to the liquid–vapor equilibrium. The reader may complete it by revisiting previous courses.

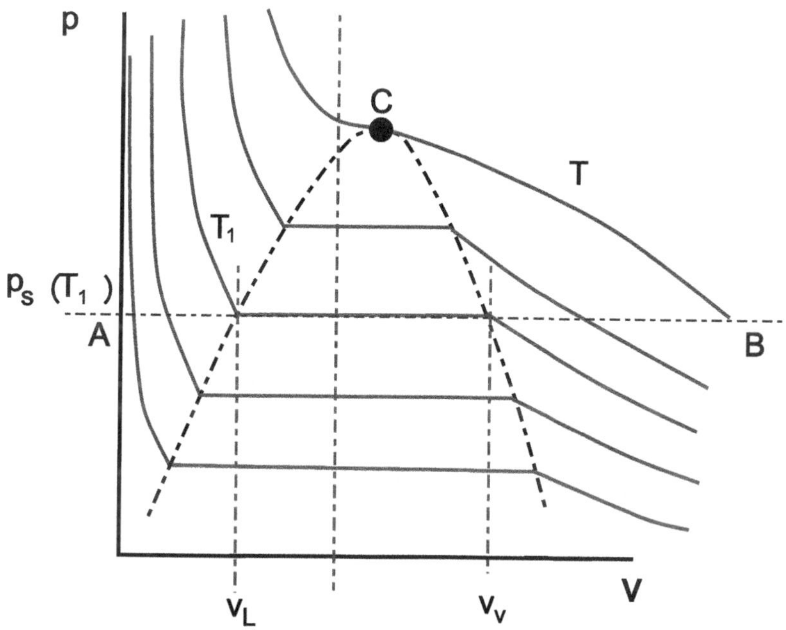

Figure 1.1. *Clapeyron diagram: Andrews' isothermals*

Let us consider a *Clapeyron diagram*, which for a pure substance represents the relation between its specific volume and its pressure.

The specific volume v, or the volume per unit mass, therefore expressed in $m^3\,kg^{-1}$ is the inverse of density ρ: $v = \dfrac{1}{\rho}$.

The **Clapeyron diagram** is composed of isothermals, which are commonly known as the *Andrews' isothermals*. It can be noted that each isothermal passing below point C reaches a plateau. All the plateaus belong to a dotted curve. The left part of this curve is known as the *boiling point curve*, while the right part is the *dew point curve.*

Each isothermal cutting the dew point curve has three parts:

– a left part, where the body is in the liquid state: this is referred to as compressed liquid;

– a right part where the body is in the gaseous state: this is referred to as vapor;

– a plateau limited by the boiling/dew point curve: on this plateau, the liquid and gaseous phases coexist; their dosage determines the specific volume. At the left end of the plateau $v = v_L$, where v_L is the specific volume of the liquid under pressure p_S and temperature T. At the right end of the plateau $v = v_V$, where v_V is the specific volume of the liquid under pressure p_S and temperature T.

The pressure of a plateau is known as the saturation vapor pressure p_S. It is a direct function of the temperature of the isothermal characterizing the plateau: $p_S = p_S(T)$.

It is possible to move along a horizontal line AB (which is therefore isobaric), for example, by modifying the volume of fluid, which is accompanied by a heat transfer, or to determine the phase change by a heat transfer, which involves a change of specific volume. From A to B, there is a succession of compressed liquid, an area of equilibrium between mixture and vapor increasingly richer in gas to the right, and then gas.

The passage of a mass m from the liquid state to the vapor state on the saturation plateau requires an amount of heat L. Expressed in C, L is the heat of vaporization (or specific enthalpy of vaporization Δh_V, as this heat is provided under constant pressure, which is the saturation vapor pressure p_S).

Clapeyron established a significant relation between the heat of vaporization, the specific volumes of the two phases and the differential of the saturation vapor pressure with respect to temperature:

$$\frac{dp_{Sat}}{dT} = \frac{L}{T(v_V - v_L)} \quad [1.1]$$

This formula can be used to calculate the heat of vaporization based on the data provided by the Clapeyron diagram.

At a point C, the isothermal is tangent to the dew point curve. In this case, the plateau is reduced to a point. C is known as the ***critical point***. The isothermals below this point correspond to a state of dense matter, a fluid that is neither liquid, nor vapor, its state being referred to as the ***supercritical state***.

This critical point is characterized by a critical pressure p_C and a critical temperature T_C, which are highly variable from one fluid to another.

Several examples are listed below:

Water: $p_C = 22.12\ MPa$ and $T_C = 374.15\ C$

Methane: $p_C = 4.6\ MPa$ and $T_C = -82.6\ C$

Carbon dioxide: $p_C = 7.29\ MPa$ and $T_C = 31.3\ C$

1.2.1.2. *Capillarity*

The presence of gases and liquids coexisting in a flow leads to the presence of interfaces. The existence and deformation of these interfaces give rise to forces, known as interface forces or capillary forces. Although weak, these forces play a first-order role in the dynamics of bubbles and pockets whose existence will be subsequently described.

Capillary forces occur as soon as an interface is stretched. These forces have their origin in molecular interactions and they compensate in volume. Each molecule is "pulled in every direction by its neighbors". At an interface, the symmetry is broken and intermolecular forces become manifest.

The fundamental law can be written as the expression of the force F needed to stretch a length l of the interface:

$$F = \sigma l \tag{1.2}$$

here, σ is the surface tension.

It is a quite weak coefficient expressed in $N.m^{-1}$. Its value for water at $T = 20\ C$ is generally considered $\sigma = 7.2 10^{-2}\ N.m^{-1}$. This value should be considered with caution. In fact, at temperature

$T = 0\ C$, its value is $\sigma = 7.564 10^{-2}\ N.m^{-1}$

and at $T = 37\ C$, it is $\sigma = 7 10^{-2}\ N.m^{-1}$

The accuracy of these figures may be misleading. They are valid for very clean water. The slightest impurity, unavoidable in many practical cases, leads to a decline of the value, which may reach $\sigma = 6.5 10^{-2}\ N.m^{-1}$ or even less in the case of notable pollution.

It is important to note that force is not proportional to the stretch, as it is for a spring. The force remains the same irrespective of stretch, until breakage. On the other hand, the force is proportional to the length of the stretched interface.

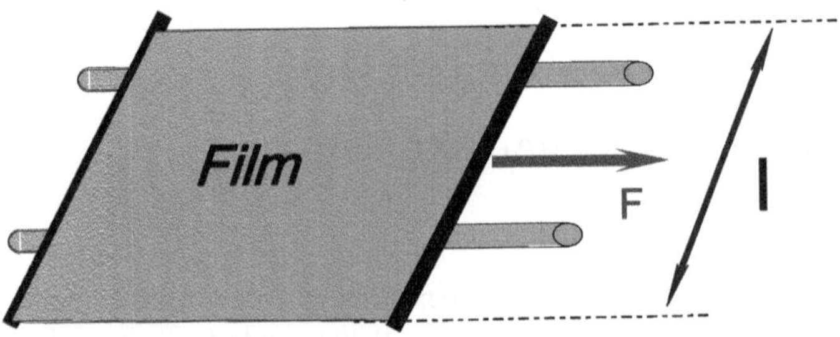

Figure 1.2. *Capillarity: stretching of a liquid film*

This is often illustrated by showing the stretching of a plane film over a length l. The force is then:

$$F = 2\sigma l \tag{1.3}$$

the coefficient 2 in this formula results from the existence of ***two*** interfaces.

In the physics of two-phase systems, interfaces of interest have a spherical geometry. In this case, a pressure difference occurs when crossing the interface. The internal pressure at the concave face is higher than the external pressure, irrespective of the fluids.

According to Laplace, in general terms, when crossing an interface, there is a pressure variation that can be written as follows:

$$\Delta p = p_i - p_e = \sigma\left(\frac{1}{R_1} + \frac{1}{R_2}\right) \qquad [1.4]$$

p_i is always the internal pressure in the concavity and $\frac{1}{R_1} + \frac{1}{R_2}$ is an invariant of a point of the surface. R_1 and R_2 are the radii of curvature on two planes perpendicular to one another.

For a spherical interface of radius R, the Laplace formula is written as:

$$\Delta p = p_i - p_e = \frac{2\sigma}{R} \qquad [1.5]$$

Three common cases can thus be considered:

– A gas bubble of radius R surrounded by liquid:

$$\Delta p = p_i - p_e = \frac{2\sigma}{R}, \qquad [1.6]$$

as there is a single interface to be crossed;

– A “free” bubble, composed of a spherical liquid thin film enclosing gas. The two gas–liquid and liquid–gas interfaces can then be assumed to have the same radius R

$$\Delta p = p_i - p_e = \frac{4\sigma}{R} \qquad [1.7]$$

as there are two interfaces to cross;

– A drop, liquid sphere of radius R

$$\Delta p = p_i - p_e = \frac{2\sigma}{R} \qquad [1.8]$$

as there is a single interface to cross.

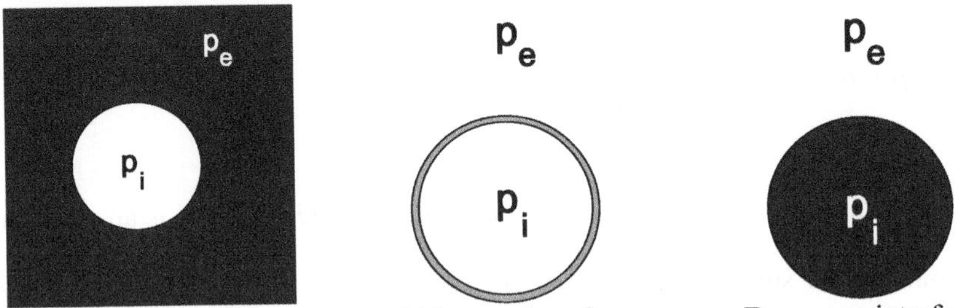

Bubble: one interface $\Delta p = p_i - p_e = \frac{2\sigma}{R}$

Bubble: two interfaces $\Delta p = p_i - p_e = \frac{4\sigma}{R}$

Drop: one interface $\Delta p = p_i - p_e = \frac{2\sigma}{R}$

Figure 1.3. *Capillarity. Drops and bubbles: various interface topologies*

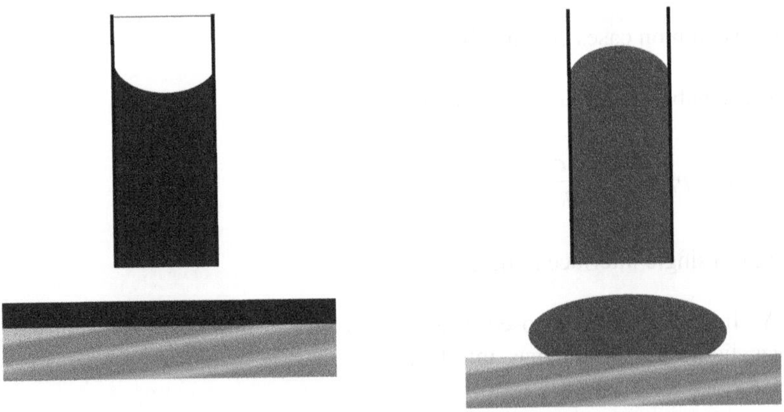

Water: concave meniscus, film

Mercury: convex meniscus, deformed drop

Figure 1.4. *Film and drop menisci*

The behavior of fluid molecules in the presence of a solid depends on the considered liquid. For example, water tends to extend over a solid. On a flat surface, this gives rise to a film, and in a tube, to a concave meniscus. On the other hand, a molten metal, such as mercury, tends to "flee" from the solid. For a drop, this leads to a contact angle, depending on the weight of the drop and the forces of attraction of the fluid surface under the drop (but the calculation of this angle may prove to be burdensome), and, in a pipe, there is a concave meniscus.

Finally, if a small diameter tube is plunged into a tank filled with water, a rise of the liquid in the tube can be noted. The height h of water above the free surface of the liquid is given by ***Jurin's law.*** When the forces (capillarity, weight of the column, pressure force on fluid interfaces) are in equilibrium, the height h is written as:

$$h = \frac{4\sigma}{\rho g D} \qquad [1.9]$$

where D is the inner diameter of the tube and σ is the surface tension constant for water–air.

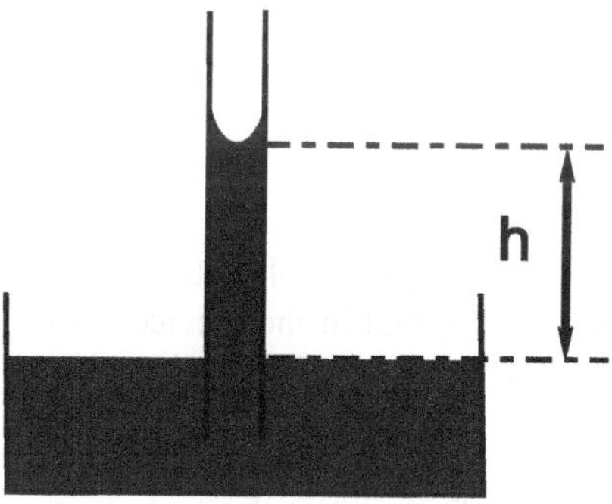

Figure 1.5. *Jurin's law* $h = \frac{4\sigma}{\rho g D}$

1.2.1.3. *Bubble dynamics*

Bubble flows are widely present among liquid gas flows. The formation of bubbles and the evolution of their size represent a wide range of significant problems in the processes.

Moreover, flows may travel through piping systems featuring cross-sectional changes or pipe accidents giving rise to significant pressure variations.

The technological impact of these problems is the object of an active fundamental research field. This quite subtle physics is generally little known, and largely ignored by the literature.

The in-depth study of this domain is beyond the scope of this book. Mentioning it here, together with its most outstanding results, appeared to be a necessity. This may motivate any interested reader to refer to the literature, most often under the title of Cavitation (see, in particular, the collective work by J.P. Franc et al. entitled *Cavitation. Physical Mechanisms and Industrial Aspects* [FRA 95]).

Two types of phenomena motivate these studies:

– the occurrence of vapor by heating of the liquid, or boiling;

– the occurrence of vapor by reduction of the pressure in the liquid, or cavitation.

The transition from liquid into vapor at the core of a liquid is more complex in practice. It involves the production, and growth of bubbles, which is the boiling phenomenon.

Bubble formation requires the temperature to reach, for a given pressure of the liquid, the saturation temperature corresponding to this temperature $T_{Sat}(p)$. In fact, the required temperature gap $\Delta T = T - T_{Sat}$ should be at least $\Delta T = 5\,C$.

The creation of a bubble in a liquid involves the existence of a seed. The seed is often a micro-volume of gas adsorbed in the crevices of the surface of the heating element (micro-cracks, etc.). The micro-bubble is thus able to grow.

Indeed, theoretically speaking, the diameter of a bubble should be zero at its creation, which would mean infinite internal pressure or, at least, for a bubble of very small diameter, an extremely high pressure. A bubble would therefore occur as a consequence of the inclusion of gas that is foreign to the vapor and partly insoluble.

Two problems arise:

– Bubble size evolution depending on external pressure stresses. This is expressed here by the Rayleigh–Plesset equation.

– Evolution of a bubble inside which the liquid vapor and a foreign gas coexist. This is expressed here by Blake's theory.

a) The dynamics of a bubble subjected to an ambient pressure (external to the bubble) that varies in time is a complex problem, in the realm of unsteady fluid mechanics, which must simultaneously take into account viscosity forces and surface tension forces. The equation describing this is known as the **Rayleigh–Plesset equation**, and sometimes as the **Besant–Rayleigh–Plesset equation.**

This equation has the form of the Navier–Stokes equation in spherical coordinates, in the hypothesis of spherical symmetry.

It is a relatively complex second-order nonlinear differential equation:

$$R\frac{d^2 R}{d\,t^2}+\frac{3}{2}\left(\frac{d\,R}{d\,t}\right)^2+\frac{4\nu_L}{R}\frac{d\,R}{d\,t}+\frac{2\sigma}{\rho_L R}+\frac{\Delta p\,(t)}{\rho_L}=0 \qquad [1.10]$$

where:

$R(t)$ is the sought-for time evolution of the radius of the bubble.

ρ_L and ν_L are respectively the density and the kinematic viscosity of the liquid in which the bubble is immersed. σ is the surface tension between the liquid and the vapor surrounding the bubble.

$\Delta\,p(t)=p_\infty-p_i$ is the pressure differential between the internal pressure in the bubble p_i, assumed to be known and $p_\infty(t)$ the pressure "far" from the bubble, which is imposed.

The resolution in $R(t)$ gives the evolution of a bubble subjected to a variation of ambient pressure.

This resolution is complex. Simplifying hypotheses are sometimes used, notably assuming that the liquid is a perfect fluid ($\nu_L=0$).

It is important to note that when $\Delta\,p$ is fixed, the radius $\Delta\,p$ no longer varies with time. The context is then that of fluid statics, and the Laplace formula is:

$$\frac{2\sigma}{\rho_L R}+\frac{\Delta p}{\rho_L}=0 \qquad [1.11]$$

or:

$$p_i - p_\infty = \frac{2\sigma}{R} \qquad [1.12]$$

b) Blake's theory (1949)

Blake was interested in the static equilibrium of a cavitation bubble. This is how he was able to explain the unstable character of this phenomenon.

Consider the existence in the liquid of a very small bubble, known as a seed, a sphere of radius r and filled with gas and vapor. Blake's innovative idea was to assume that this very small vapor bubble is polluted by a gaseous impurity.

Consider a seed containing liquid vapor at pressure p_V and a foreign gas at pressure P_G.

At equilibrium, the following can be written as:

$$p_i - p_\infty = \frac{2\sigma}{R} \qquad [1.13]$$

where:

$$p_i = p_G + p_V \qquad [1.14]$$

or:

$$p_\infty = p_G + p_V - \frac{2\sigma}{\rho_L R} \qquad [1.15]$$

where pressure p_V is a saturation vapor pressure. Assuming that temperature is constant, it is fixed.

On the other hand, p_G varies with volume and therefore with the radius R of the bubble.

Consider R_0, the radius of the bubble at the beginning of a process, and p_{G0}, the pressure of the gas for this radius. For another radius R, the following can be written as:

$$\frac{p_G}{p_{G0}} = \left(\frac{R_0}{R}\right)^3 \qquad [1.16]$$

which is the expression of Boyle–Mariotte's law under constant temperature.

The relation between pressure p_∞ and radius $R(p_\infty)$ is therefore given by:

$$p_\infty = p_{G0}\left(\frac{R_0}{R}\right)^3 + p_V - \frac{2\sigma}{R} \qquad [1.17]$$

When R increases, this curve is asymptotic to $p_\infty = p_V$:

– this can be deduced from:

$$p_\infty = p_{G0}\left(\frac{R_0}{R}\right)^3 + p_V - \frac{2\sigma}{\rho_L R} \text{ when } R \text{ tends to infinity;} \qquad [1.18]$$

– the curve of variation of p_∞ with R depends on the rate of foreign gas present in the bubble. For a given value of p_{G0}, this curve descends to a minimum, referred to as critical.

A critical value of radius R_C is thus obtained:

$$R_C = \sqrt{\frac{3\,p_{G0}\,R_0^3}{2\sigma}} \qquad [1.19]$$

and a critical value of the pressure:

$$p_C = p_{G0}\left(\frac{R_0}{R_C}\right)^3 + p_V - \frac{2\sigma}{\rho_L R_C} = p_{G0}\frac{2\sigma R_0^3}{3\,p_{G0} R_0^3} + p_V - \frac{2\sigma}{R_C}. \qquad [1.20]$$

or:

$$p_C = p_V - \frac{4\sigma}{3R_C} \quad [1.21]$$

Below this pressure, and above this value of the radius, the radius of the bubble cannot be given by the static solution. Depending on the experimental conditions, a bubble explosion is generally possible.

1.2.1.4. *Nucleate boiling*

In Andrews' experiments, when pressure conditions at constant temperature change, a liquid–vapor equilibrium can be reached without difficulty in practice. This equilibrium results from an evaporation or condensation at an interface in a piston–cylinder system.

The evaporation or condensation at an interface is one of the ways in which the passage between liquid and vapor states occurs in practical systems. This occurs at the free surface of films.

This bubble generation is known as nucleate boiling. It is an essential mechanism in the two-phase heat transfer through pipes. As already noted, a minimum requirement for boiling is the saturation temperature. In a pipe subjected to a heat flux, a thermal gradient is established in the fluid, temperature decreasing starting from the wall. When a thermal regime sets in, the core of the fluid may remain at a temperature below a temperature gap $\Delta T = T - T_{Sat}$ while this temperature may exceed a temperature gap $\Delta T = T - T_{Sat}$ in the area near this wall. Nucleate bubbles may then form at the wall. This is referred to as *subcooled nucleate boiling*. If all the liquid has a temperature equal to or higher than a temperature gap $\Delta T = T - T_{Sat}$, this is referred to as *saturated nucleate boiling*.

Our objective is to present here an empirical approach to the heat transfer in a vertical tube. However, for a better understanding of the problems involved by this transfer, the results of the study of heat transfer during *pool boiling* are presented.

Consider a setup in which a volume of liquid is in a container. This liquid is heated, either through a wall, or in practice by an electrical heating element (immersion heater). The process is similar in practice to that occurring in a pressure cooker or a domestic electric water heater. It is important to note that interesting phenomena occur at the surface of the heating source.

Based on this already quite old experiment conducted by Nukiyama [NUK 66], the differential of the wall temperature of the heating element T_W and of the saturation temperature $\Delta T = T_W - T_{Sat}$ is represented in the abscissa. On the ordinate, there are the heat flux density T_W transmitted to the fluid and the convection coefficient $h = \frac{\varphi_W}{T_W - T_O}$, where T_O is the fluid core temperature.

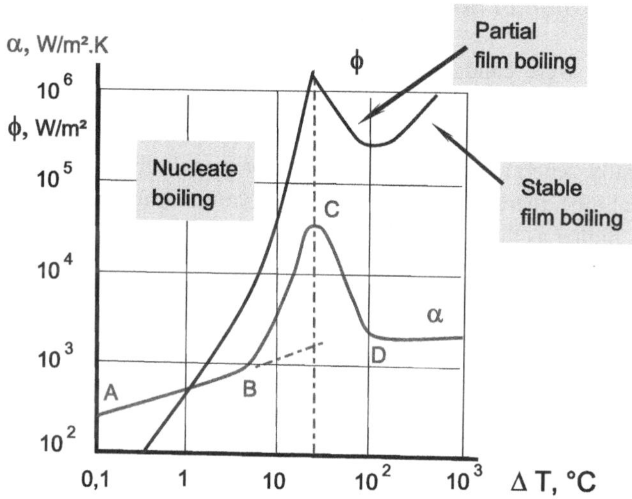

Figure 1.6. *Heat transfer in pool boiling*

The curve of the convection coefficient can serve as a guide. It is important to note the presence of several zones, therefore heating regimes, varying with the temperature differential controlled by the wall temperature:

From **A** to **B**, there is natural convection in a single phase liquid.

From **B** to **C**, there is nucleate boiling. As noted above, this point is situated at a differential $\Delta T = 5\,C$.

In **C**, there is a transfer peak. At this point, steam generation is such that it destabilizes the liquid film: the solid interface is partly covered by a vapor film. The convection coefficient and consequently the flux are limited by this vapor film whose thermal resistance is higher than a liquid convection coefficient.

An unstable liquid film can be noted from **C** to **D**, leading to a fluctuating contact between the wall and the gas and the liquid. As the proportion of vapor increases, the flux decreases with temperature.

Finally in **D**, a continuous vapor film is formed, the convection coefficient is stabilized and the flux can then increase again with the temperature differential.

1.2.2. *Classification of flows*

Many studies were conducted without heating in "two-phase loops" where the liquid and gas flow rates are controlled. Flow quality is thus imposed as an experimental parameter. This research method can be used to reproduce flow types that are naturally generated along a vertical or horizontal tube, when it is heated. In a heated vertical tube, there can be a succession of various characteristic types of flow, whose rate is higher as the heating gets stronger.

Ascending vertical flows

– Bubbly flow. Bubbles can be independent or gathered in clusters.

– Plug flow or slug flow, where the gas plugs are carried by the flow. It can be noted that these plugs are rounded at the top and rather flat at the bottom.

– Annular flow. The liquid forms a film along the wall. A continuous central flow of air extracts drops from this film. Several transition phases can be noted, a quite chaotic *churn* flow, then a film flow and a *dispersed flow*, in which the drops sometimes gather in "clusters" and is referred to as *wispy flow*.

Horizontal flows

When the gas flow rate increases, the following types of flows can be successively noted in a horizontal tube:

– A bubbly flow. The gas is contained in the liquid flow body in the form of bubbles.

– A slug flow or plug flow in which bubbles have coalesced and given rise to slugs.

– A stratified flow, where the flows of gas and liquid are continuous and superimposed. At this stage, the interface is calm.

– A wavy flow. The speed difference between gas and liquid deforms the interface and gives rise to waves.

– A slug flow. The amplitude of waves is sufficient to reach the upper part of the tube and slugs are once again formed.

– An annular flow. An annular liquid film is crossed by a continuous flow of gas. The drops extracted from the film by instability form a cloud at the center of the gas flow.

A descending vertical flow features aspects that have already been mentioned with the ascending tube, with several variants:

– bubbly flow;

– slugs;

– falling film: when the flow rates are low, the liquid descends along the wall in a film. Obviously, this configuration cannot be observed in ascending flow, where the only way the film can be generated is by the effect of the viscosity of the gas flow;

– bubbly falling film;

– churn flow;

– dispersed annular flow.

Typology of two-phase flows

In two-phase flow, the geometry of the interfaces between vapor and liquid comes under different forms (surfaces of films, bubbles, slugs). This leads us to determine various typologies, also referred to as two-phase flow topologies. Only pipe flows are covered here. Given the large difference in density between vapor and liquid (typically from 10 to 10^3 depending on the pressure in a section), gravity has a strong influence and requires a distinction being made between ascending and descending vertical flows, as well as horizontal flows. Annular tubes and spaces should also be distinguished.

Although all geometries may be present in real heat exchangers, this book focuses on ascending tubes. However, for the reader's information, indications will be given on descending and horizontal flows.

The motivation for the numerous experiments conducted in order to identify these flows is given by the necessity to identify, classify, and predict the occurrence of flowing typologies. They are also an experimental support for the modeling and prediction of the mechanics of fluids and two-phase transfers.

As already mentioned, for any reader with a desire to specialize in this domain, this chapter can only be an introduction to a more complete literature.

Parameters

The parameters determining a flow that are most commonly indicated in the publications are the following:

Mass flow rate of the liquid: q_{mL}

Mass flow rate of the vapor: q_{mV}

(Actual) mass quality

$$x = \frac{q_{mV}}{q_{mV} + q_{mL}} \quad [1.22]$$

Volume flow rate of the liquid: q_{VL}

Volume flow rate of the vapor: q_{VV}

Volume quality

$$\beta = \frac{q_{VV}}{q_{VV} + q_{VL}} \quad [1.23]$$

Surface rates

$$J_V = \frac{q_{mV}}{\rho_V \, S} \quad [1.24]$$

$$J_L = \frac{q_{mL}}{\rho_L \, S} \quad [1.25]$$

"Mass rates" are also considered:

$$\dot{m}_V = \frac{q_{mV}}{S} = \rho_V \, J_V \quad [1.26]$$

$$\dot{m}_L = \frac{q_{mV}}{S} = \rho_L J_L \qquad [1.27]$$

The following expression can be written as:

$$q_m = q_{mV} + q_{mL} \qquad [1.28]$$

hence:

$$x = \frac{q_{mV}}{q_m} \qquad [1.29]$$

and:

$$q_{mV} = x\, q_m \qquad [1.30]$$

which leads to:

$$J_V = x_V \frac{q_m}{\rho_V} \qquad [1.31]$$

Ascending vertical flow

Various topologies are represented in Figure 1.7.

The best known flow is the *bubbly flow*. When their density increases, upon cold increase of the initial gas flow rate, or they appear during heating in the presence of heat transfer, these bubbles interact strongly and a *packed bubbly flow* can be noted.

Due to their interaction under high density conditions, bubbles coalesce and a *slug flow* can be noted. At this level, there is a distinction between plug flow and slug flow. Plug designates here a volume, a gas pocket. It appears that putting several plugs end to end results in a long pocket whose oscillating behavior recalls the advancement of a slug.

Increasing the vapor rate flow leads to a film flow; this takes place through a chaotic destabilization leading to a *churn flow* configuration. An annular flow is thus obtained.

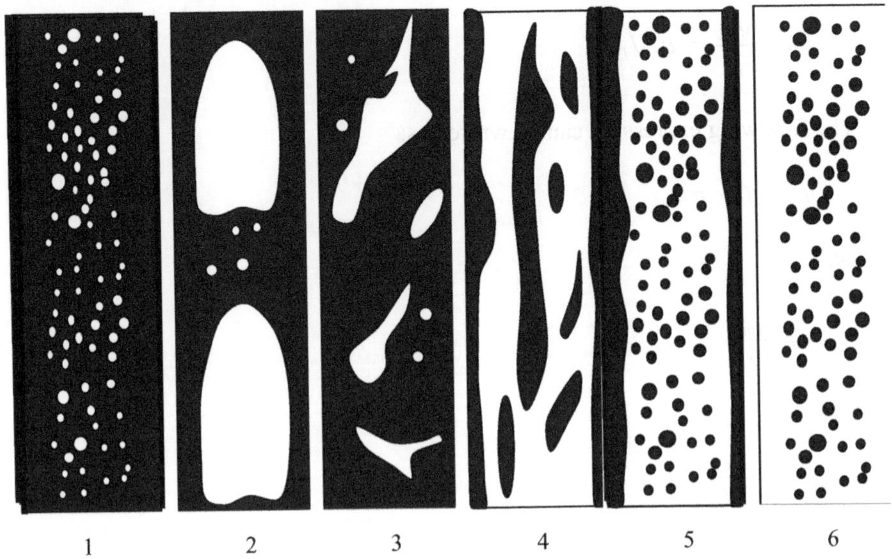

1) Bubbly flow

2) Slug flow

3) Churn flow (shaked)

4) Churn flow (chaotic)

5) Bubbly film

6) Dispersed annular flow

Figure 1.7. *Ascending vertical flow*

Descending vertical flows

These flows are represented in Figure 1.8.

Bubbly flows and slug flows occur, and a film flow emerges. This is a falling film this time.

This falling film evolves into a bubbly film. Due to churn, it may evolve into a dispersed annular flow. In this case, the velocity differential between vapor and liquid allows a surface sputtering and the gas flow carries a mist.

It is important to note a particular phenomenon, which occurs when an ascending vapor flow emerges. If the velocity of this flow is sufficient, the gas–liquid interfacial friction is significant and tends to bring up the film. This phenomenon, which may be annoying in the installations, is referred to as "*flooding*".

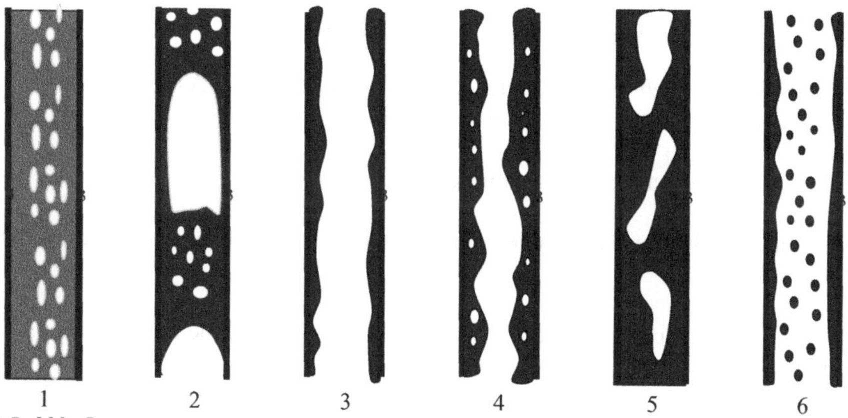

1) Bubbly flow

2) Slug flow

3) Falling film

4) Bubbly film

5) Churn flow

6) Dispersed annular flow

Figure 1.8. *Descending vertical flow*

Horizontal flow

Various topologies are presented in Figure 1.9.

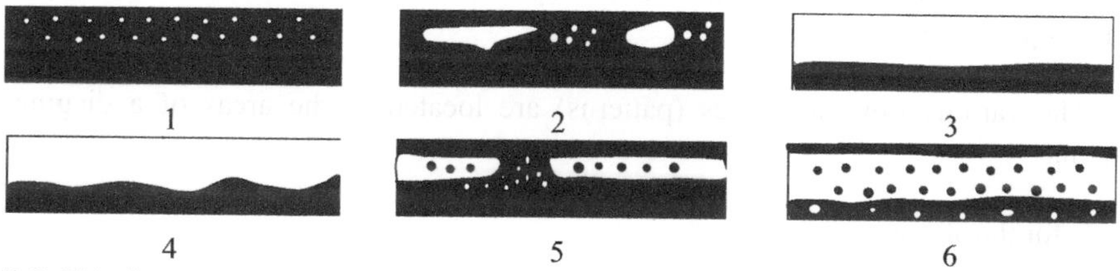

1) Bubbly flow

2) Plug flow

3) Stratified flow

4) Wavy flow

5) Slug flow

6) Annular flow

Figure 1.9. *Horizontal flow*

Due to gravity, there is a certain stratification of flows.

First, there are *bubbly flows* and a *plug flow* carrying slugs. As can be noted, the gaseous phases tend to concentrate on the upper part of the tube.

From the same perspective, a stratified flow can be noted.

For quite high values of gas velocities, the interface becomes undulated. This is the *wavy flow*.

Since the amplitude of waves increases with gas velocity, the crest of the waves reaches the upper wall of the tube, and a *slug flow* emerges.

This flow may evolve into an annular flow, where the extraction of drops leads to a mist carried by the gaseous phase.

Validity areas of these flows

Identification works, relying on visualizations, are doubled given the necessity to predict the occurrence of these various typologies. The literature on this subject is quite rich, so only references will be provided here. Focusing on the vertical pipe flows, one of the methods for predicting the occurrence is presented here. Similar to the others, it relies on a diagram for positioning the experimental operating point of a flow on a diagram. The parameters on these scales vary from one author to another.

The Hewitt and Roberts method [HEW 69] is presented here. It relies on experiments conducted on the air/water couple up to 59 bar, and water vapor/liquid water up to 69 bar.

The various flow structures (patterns) are located in the areas of a diagram having:

– for the abscissa:

$$\rho_L J_L^2 = \frac{\bar{G}^2(1-\beta)^2}{\rho_L} \qquad [1.32]$$

– for ordinate:

$$\rho_G J_G^2 = \frac{\bar{G}\beta^2}{\rho_G} \qquad [1.33]$$

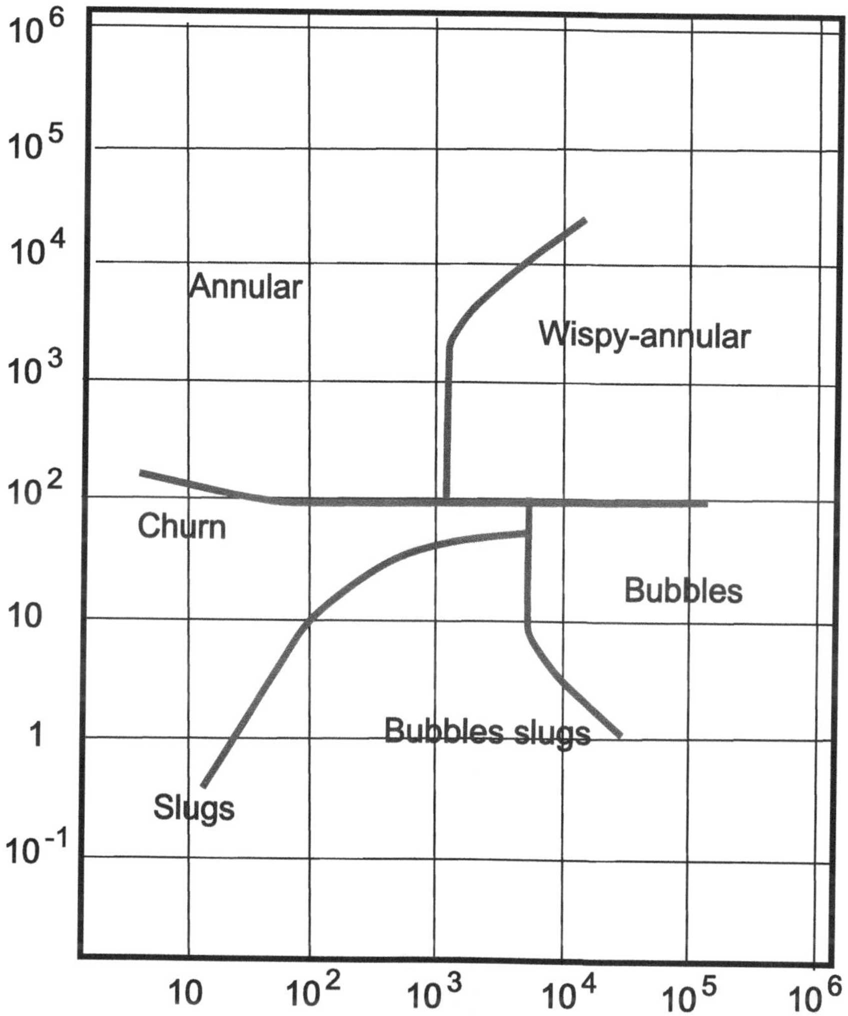

Figure 1.10. *Diagram* in the abscissa: $\rho_L J_L^{\,2} = \dfrac{\bar{G}^2(1-\beta)^2}{\rho_L}$;

in the ordinate $\rho_G J_G^{\,2} = \dfrac{\bar{G}\beta^2}{\rho_G}$

1.2.3. *Topologies of a heated vertical flow*

Various types of flow that may occur in a vertical tube have been described. As already mentioned, in fundamental studies, the flow is most often obtained by controlling the quality (ratio of mass flow rates of gas and liquid). The gas is most often air, distinct from the liquid vapor.

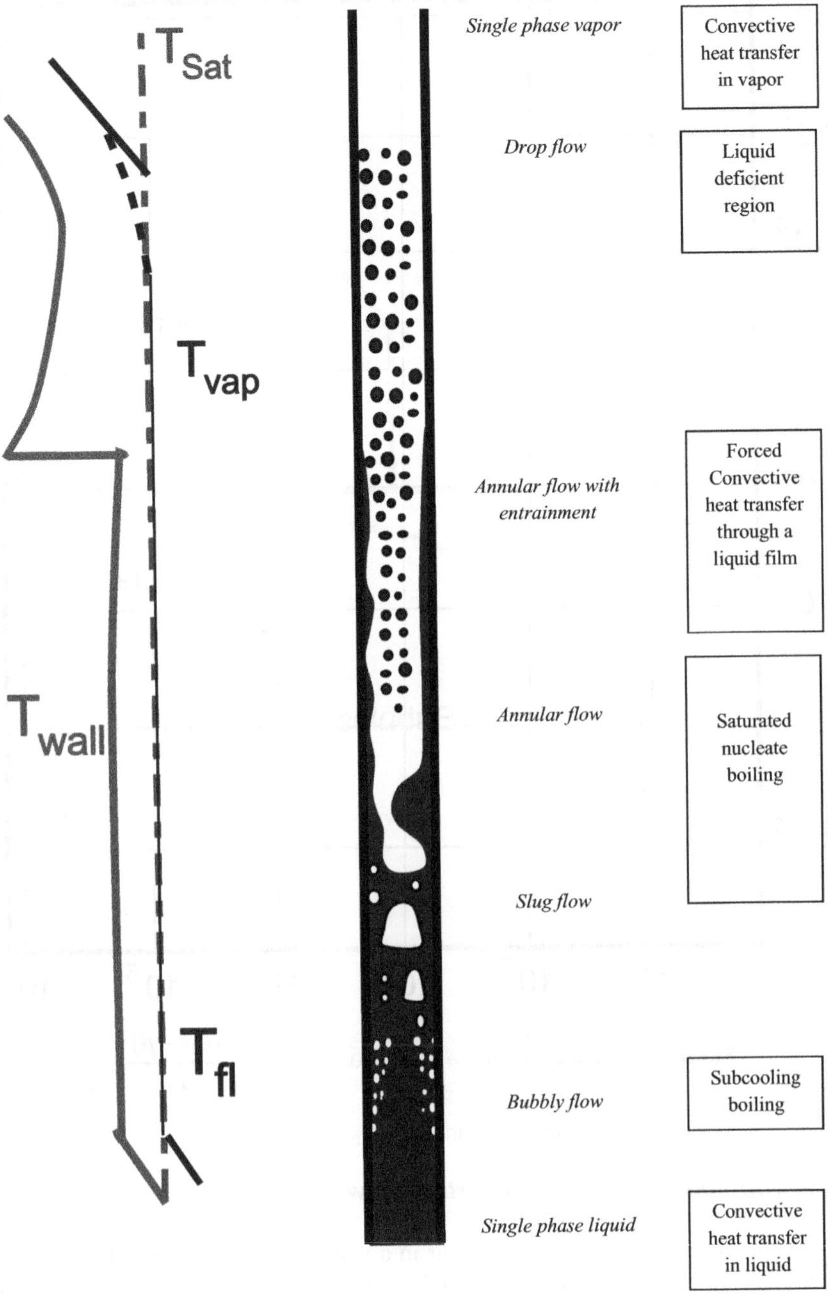

Figure 1.11. *Ascending vertical flow. Flow topologies and transfer zones*

In the case of a heated tube through which there is an initially liquid, therefore single phase, flow, the various phases of the flow and the various structures (flow patterns) of this flow are determined by the evolution, the history of the heating. Flow structures follow one another from the bottom to the top of the tube, all of them being interdependent, in the "history" of the heated flow.

Not all the structures are present, their occurrence depending on the length of the tube and the imposed heat load.

They are presented using a schematic representation that is quite commonly present in one form or another in various treatises.

The case considered here is that of a wall transfer at constant flux density φ_W .

The occurring flow structures (patterns) are successively analyzed, followed by the types of heat transfer involved, and finally a schematic representation of the forms taken by the distribution of wall and core temperature is provided.

From the bottom to the top of the tube

A single phase zone (liquid flow) remains at a temperature below the saturation temperature T_{Sat} (temperature of the liquid–vapor equilibrium plateau of Andrews' isothermals). It is important to note that the effect of heating is a parallel increase of the wall and core temperatures. As long as no point of the liquid reaches the temperature T_{Sat} , the flow remains single phase.

A nucleate boiling zone:

– given the thermal gradient controlling the transfer, the temperature T_{Sat} will be reached close to the wall before the axis and the bubbles will form at the wall. In this first zone, the density of the bubble remains moderate and localized near the wall, with a migration;

– when the temperature reaches (and exceeds) T_{Sat} throughout a section, there is bubbly flow throughout the tube.

Slug flow: as the density of bubbles increases, their interactions are strong, and they then coalesce into slugs.

Annular flow: as they grow, these slugs also coalesce and they generate a liquid film flow surrounding an inner gas flow. For mass conservation reasons, the inner velocity of the gas exceeds that of the film (slowed down, among others, by the viscosity at the wall) and this film is referred to as entrained. The velocity difference between vapor and liquid also leads to a destabilization of the interface (Kelvin–Helmholtz instability) and the formation of a cloud of drops carried by the gaseous flow.

Drop flow: at the interface of the film, there is skimming by atomization and also evaporation. The liquid rate flow of this film will therefore decrease, the film thickness will decrease from bottom to top, until they disappear. There only remains a drop flow that evaporates until it becomes a single-phase flow, this time a vapor flow.

The disappearance of the film is, both literally and figuratively, a catastrophe. Indeed, it leads to a sudden discontinuity of the convection coefficient T_{Sat}, which decreases drastically. The flux being conserved, the temperature differential strongly increases (the schematic representation is resented further). This differential is reflected by the wall temperature, whose integrity may be compromised.

This phenomenon is known as *dry out*. The consequences in various industrial installations, particularly in the nuclear field, can only be imagined.

This is designated by terms such as boiling crisis, critical heat flux (CHF), departure from nucleate boiling (DNB) or burnout (BO).

From a thermal point of view, various transfer zones can be qualified based on these flow structures:

– zone of convective heat transfer between the wall and the liquid;

– zone of subcooled boiling corresponding to wall boiling;

– zone of saturated nucleate boiling, joining the developed bubbly flow and the slug flow zone;

– zone of forced convective heat transfer through the liquid film;

– zone of flow in the absence of liquid, joining evaporating drop flow and pure vapor.

Temperature profiles

A general form of temperature profiles can be logically deduced.

Recalling that flux density is constant: the zone of liquid convection leads to a parallel increasing evolution of the wall and fluid temperatures.

Since the occurrence of saturated nucleate boiling, the fluid temperature remains constant and equal to T_{Sat}. In the annular and film zones, the decrease of the film leads to a decrease of the thermal resistance. The wall temperature slightly decreases.

At dry out, the drastic decrease of the convection coefficient leads to an increase of thermal resistance and, the vapor remaining close to T_{Sat}, to a sudden and significant increase in the wall temperature.

Finally, the return to a two-phase flow leads to a parallel evolution of the wall and core temperatures.

Definition of quality

Given the mass flow rates of the vapor and the liquid, denoted by q_{mV} and q_{mL} respectively, in a tube section, the quality x is defined by:

$$x = \frac{q_{mV}}{q_{mV} + q_{mL}} \qquad [1.34]$$

It can also be shown that:

$$x = \frac{h(z) - h_L}{L} \qquad [1.35]$$

where L, h_V and h_L are respectively the heat (enthalpy) of vaporization and the specific enthalpies of vapor and liquid; $h(z)$ is the specific enthalpy in the considered section, of altitude z.

1.2.4. *Calculation of fluxes*

Each flow structure leads to a different formulation for the calculation of wall flux. The literature on this subject is quite complex. A selection is proposed here.

Single-phase flow

$$\Delta T = T_W - T_L(z) \qquad [1.36]$$

$$\varphi_W(z) = h_{L0}\left[T_W - T_L(z)\right] \qquad [1.37]$$

Dittus and Boelter

$$Nu_D(z) = \frac{h_{L0}(z)D}{\lambda_L} = 0.023\, R_D^{0.8}\ \mathrm{Pr}_L^{\frac{1}{3}} \quad [1.38]$$

where R_D is the Reynolds number of the liquid flow:

$$R_D = \frac{V_{qL}\, D}{\nu_L} = \frac{\dot{m}_L\, D}{\mu_L} = \frac{q_{mL}\, D}{S\,\mu_L} \quad [1.39]$$

where S is the cross-section of the tube and $\dot{m}_L$ is the "mass rate" of the liquid

$$\dot{m}_L = \rho_L\, V_{qL} = \frac{q_{mL}}{S} \quad [1.40]$$

and the Prandtl number of the liquid:

$$\mathrm{Pr}_L = \frac{\nu_L}{a_L} = \frac{\mu_L\, c_{pL}}{\lambda_L} \quad [1.41]$$

The Dittus–Boelter formula does not take into account the viscosity variation due to the temperature on a section.

The Sieder–Tate formula may be preferred, which calculates the Nusselt number by:

$$Nu_D = \frac{h_L\, D}{\lambda_L} = 0.023\ \mathrm{Pr}^{\frac{1}{3}}\ R_D^{0.8}\left(\frac{\mu}{\mu_W}\right)^{0.14} \quad [1.42]$$

which leads to

$$h_L = 0.023\frac{\lambda_L}{D}\ \mathrm{Pr}^{\frac{1}{3}}\ R_D^{0.8}\left(\frac{\mu}{\mu_W}\right)^{0.14} \quad [1.43]$$

The two formulas differ only by the term $\left(\frac{\mu}{\mu_W}\right)^{0.14}$

The use of this term leads to an iterative process in the calculation of the variation of the axial temperature. It is important to note, for example, that at $p = 15\,bar$ for respective temperatures at the wall and the core $T_W = 200\,C$ and $T_L = 100\,C$, the ratio is $\dfrac{\mu}{\mu_W} = \dfrac{1.39.10^{-4}}{2.82.10^{-4}}$ and the term $\left(\dfrac{\mu}{\mu_W}\right)^{0.14}$ is equal to 0.905 which is a 10% gap between the two formulas.

Subcooled nucleate boiling

ONB = Onset of boiling

Bergles and Rosenhow [BER 64]

$$\left(T_W - T_{Sat}\right)_{ONB} = 0.556\left(\frac{\varphi_W}{1082\,p^{1.156}}\right)^{0.463\,p^{0.0334}} \qquad [1.44]$$

David and Anderson [DAV 66]

$$\left(T_W - T_{Sat}\right)_{ONB} = \left(\frac{8\sigma\,\varphi_W\,T_{Sat}}{\lambda\,L\,\rho_V}\right)^{0.5} \qquad [1.45]$$

To extend this to liquids other than water, Frost and Dzacovic propose the multiplication of this formula by the Prandtl number of the liquid:

$$\left(T_W - T_{Sat}\right)_{ONB} = \left(\frac{8\sigma\,\varphi_W\,T_{Sat}}{\lambda\,L\,\rho_V}\right)^{0.5} \mathrm{Pr}_L \qquad [1.46]$$

The transfer is then calculated with the same convection coefficient as for pure liquid, but the temperature difference is recalculated in an adapted manner:

$$\varphi_W(z) = h_{LO}\left[T_W - T_L(z)\right] = h_{LO}\left[\left(T_W - T_{Sat}\right)_{ONB} + \left(T_{SAT} - T_L(z)\right)\right] \qquad [1.47]$$

Fully developed subcooled boiling

The wall flux density (heat load) is imposed.

The wall temperature is expressed as a function of the wall flux density and pressure:

$$T_W(z) = T_W(\varphi_W, p) \qquad [1.48]$$

Rosenhow [ROS 52]:

$$\frac{c_p \, \Delta T_{Sat}}{L} = C_{FSB} \left[\frac{\varphi_W}{\mu_L \, L} \left(\frac{\sigma}{(\rho_L - \rho_V) g} \right)^{0.5} \right]^{0.33} \mathrm{Pr}_L^{1.7} \qquad [1.49]$$

C_{FSB} is a constant that varies according to the authors, depending on the experimental data used to establish a correlation.

The presence of the surface tension can be noted, which highlights the influence of nucleation in the physical process. Similarly, the constant varies depending on the nature of the metallic material of the wall.

Without pretending to be exhaustive, an example is provided below for a heat transfer with water:

Geometry	**Nature of the tube**	C_{FSB}
Vertical tube $D = 4.56\ mm$	Stainless steel	0.06
Vertical tube $D = 27.1\ mm$	Copper	0.013
Horizontal tube $D = 14.9\ mm$	Stainless steel	0.015
Horizontal tube $D = 2.39\ mm$	Stainless steel	0.020

Jen and Lottes [JEN 51] for water:

$$\Delta T_{Sat} = 25 \, \rho_W^{0.25} \exp \frac{-p}{62} \qquad [1.50]$$

(p in atmospheres), modified by Thom:

$$\Delta T_{Sat} = 22.65 \, \rho_W^{0.5} \exp \frac{-p}{87} \qquad [1.51]$$

valid for the ranges $115-340\ C$, $7-172\ bar$ and $11<\dot{m}<1.05.10^4\ kg.m^{-2}.s^{-2}$.

Saturated nucleate boiling

Consider simply

$$T_L(z)=T_{Sat} \tag{1.52}$$

The wall temperature becomes independent of $\dot{m}$ and x.

The limit between *saturated nucleate boiling* and the *two-phase forced convective region* is determined by a critical value of the wall flux density

$$\varphi_{W,ONB}=\frac{8h_{TP}^2\,\sigma T_{Sat}}{\lambda_L\,L\,\rho_V} \tag{1.53}$$

Here, h_{TP} is the convection coefficient in the two-phase forced convective region.

Two-phase forced convection region

The flow takes the form of an annular film

$$\frac{h_{TP}}{h_{L0}}=f\,\frac{1}{X_{tt}} \tag{1.54}$$

Here, X_{tt} is the Martinelli coefficient for turbulent flows of liquid and gas (double index tt).

Related to the head losses $X_{tt}=\dfrac{\Delta p_L}{\Delta p_V}$

X_{tt} can be calculated by:

$$X_{tt}=\left(\frac{1-x}{x}\right)^{0.9}\left(\frac{\rho_V}{\rho_L}\right)^{0.5}\left(\frac{\mu_L}{\mu_V}\right)^{0.1} \tag{1.55}$$

Forms of the function:

Dengler [DEN 56]:

$$\frac{h_{TP}}{h_{L0}} = 3.5\left(\frac{1}{X_{tt}}\right)^{0.5} \quad [1.56]$$

Guerrieri and Talty [GUE 56]:

$$\frac{h_{TP}}{h_{L0}} = 3.4\left(\frac{1}{X_{tt}}\right)^{0.45} \quad [1.57]$$

Bennett [BEN 61]:

$$\frac{h_{TP}}{h_{L0}} = 0.564\left(\frac{1}{X_{tt}}\right)^{0.74} \varphi_W^{0.11} \quad [1.58]$$

Schrock and Grossman [SCH 62]:

$$\frac{h_{TP}}{h_{L0}} = \gamma\left[\frac{\varphi_W}{\dot{m}L} + m\left(\frac{1}{X_{tt}}\right)^n\right] \quad [1.59]$$

with

$$\gamma = 7.39.10^3$$

$$m = 1.5.10^{-4}$$

$$n = 0.66$$

values of Wright, suggested by Collier:

$$\gamma = 6.70.10^3$$

$$m = 3.5.10^{-4}$$

$$n = 0.66$$

Other approaches

The following formulas are valid for the zones of saturated nucleate boiling and the two-phase forced convection region.

Chen:

$$h_{TP} = h_{NCB} + h_C$$

where h_{NCB} is the convection coefficient calculated for nucleate boiling and h_C is given by:

$$h_C = 0.023 \left[\frac{\dot{m}(1-x)D}{\mu_L} \right]^{0.8} \mathrm{Pr}_L^{0.4} \frac{\lambda_L}{D} F(X_{tt}) \qquad [1.60]$$

This recalls the formulas for single-phase convective heat transfer, with a modified Reynolds number and a corrective function $F(X_{tt})$.

$F(X_{tt})$ is given in the form of a curve that can be replaced by the expression:

– for $X_{tt} < 0.1 \ : \ F(X_{tt}) = 1$ [1.61]

– for $X_{tt} > 0.1 \ : \ F(X_{tt}) = 2.35 \left(\frac{1}{X_{tt}} + 0.213 \right)^{0.736}$ [1.62]

Forster and Zuber proposed the calculation of h_{NCB} by:

$$h_{NCB} = 1.22.10^{-3} \frac{\lambda_L^{0.79} c_{pL}^{0.45} \rho_L^{0.49}}{\sigma^{0.5} \mu_L^{0.29} L \rho_V^{0.24}} \Delta T_{Sat}^{0.24} \Delta p_{Sat}^{0.75} S \qquad [1.63]$$

where S is given by a curve that can be expressed by:

$$S = \left(1 + 2.53.10^{-6} R_{eL}^{1.17}\right)^{-1} \qquad [1.64]$$

Post dry out

Groeneveld

$$Nu_V = a\left[\mathrm{R}_{eV}\left[x+\frac{\rho_V}{\rho_L}(1-x)\right]\right]^b \mathrm{Pr}_{V,W}^c\, Y^d \qquad [1.65]$$

with

$$Y = 1-0,1\left(\frac{\rho_L}{\rho_V}-1\right)^{0.4}(1-x)^{0.4} \qquad [1.66]$$

then:

$a = 1.09.10^{-3}$

$b = 0.985$

$c = 1.41$

$d = -1.15$

Slaghterback [SLA 71]:

$$Nu_V = a\left[\mathrm{R}_{eV}\left[x+\frac{\rho_V}{\rho_L}(1-x)\right]\right]^b \mathrm{Pr}_{V,W}^c\, \varphi_W^d \left(\frac{\lambda_V}{\lambda_{V,BO}}\right)^e \qquad [1.67]$$

It is important to note an uncommon expression, where the Nusselt number depends on the flux density, which is also part of its definition.

Here, $\lambda_{V,BO}$ is the heat conductibility at "burn out" and

$a = 1.16.10^{-3}$,

$b = 0.838$,

$c = 1.81$,

$$d = 0.278\ ,$$

$$e = 0.508\ .$$

1.3. Dispersed phase flows: mists and sprays

In the broad sense, two-phase flows can be divided into two categories: pipe flows, bubbly flows, slug flows or a more complex combination of film and drop flows with dispersed phase: mists and sprays

1.3.1. *Mist*

A mist is a suspension in a gaseous fluid, possibly in a moderate flow of a cloud of drops or droplets of moderate density.

If the liquid phase and possibly the gaseous phase have a strong impulse, then it is a spray. The second category is of interest here.

Sprays have a particular place in the field of fluid mechanics and heat engineering.

In the absence of heat incidence, they relate to fluid mechanics, and are the subject of various chapters:

– the theory of instabilities for their formation;

– the particular transport by a flow, which is most often turbulent;

– possibly the theory of jets, if the cloud impulse is sufficient to drive a significant flux of air.

The presence of thermal phenomena can take various forms, a non-exhaustive list of which is as follows:

– Spray vaporization; a drop may evaporate in a calm gaseous atmosphere, or in a gaseous flow with relative velocity. The convective heat transfer between the gas and the drop is then considered in the calculation.

– Combustion of drops, distinguishing between the combustion of the individual drop and the combustion of an oxidized mixture resulting from the prevaporization of spray.

– Interaction of the spray with a solid surface: spray cooling phenomena, for example, metal quenching.

Finally, it is important to note that in certain topologies of tube flow, there are zones of mist, their formation requiring a destabilization of the liquid–gas interface of a film.

Sprays are quite common in daily life and in the industrial field. In the absence of heat engineering, there are cosmetic and pharmaceutical applications. In the industrial world, spray is important in combustion phenomena, being present in the field of industrial combustion, internal combustion in cars, aircraft and spaceship engines.

Sputtering is in fact a must for burning liquids. Combustion is a phenomenon occurring exclusively in gaseous phase. Any liquid needs to be vaporized to undergo the chemical process of combustion. The burnt quantity is proportional to a volume, and the evaporated quantity is proportional to a surface. The division into droplets optimizes this vaporization by increasing the ratio of the evaporation surface S to the volume V to be burnt; for a drop of diameter d :

$$S = \pi \frac{d^2}{4} \tag{1.68}$$

$$V = \pi \frac{d^3}{6} \tag{1.69}$$

$$\frac{S}{V} = \frac{\pi d^2}{4} \frac{6}{\pi d^3} = \frac{3}{2d} \tag{1.70}$$

It can be seen that the ratio $\frac{S}{V}$ is in $\frac{1}{d}$; it is therefore optimized for the small droplets. Moreover, it can be shown that a combustion time of the drops is proportional to d^2.

This explains why the purpose in combustion is to find the smallest possible size of the drops.

This subject cannot be extensively covered within the limited space of this book, but it seemed necessary to give the reader a minimal amount of information on a subject that has little attention paid to it in classical manuals.

The following section provides several notions on:

– the formation of sprays, the processes involved and the sputtering techniques;

– the characterization of a spray, or particle sizing;

– spray vaporization.

1.3.2. *Spray generation*

Spray generation involves a significant number of various devices, whose structure often depends on the industrial field concerned. A non-exhaustive list includes:

– car, aircraft and spaceship engine manufacturers;

– agriculture, for spreading insecticides;

– surface coverage technologies: paintings, surface treatments, etc.

The needs of these various domains differ. In order to optimize combustion, engine manufacturers are interested in decreasing vaporization times. Sprays of the smallest possible size are sought. Agriculture is most often interested in larger drops, which are less carried by the wind and offer a better coverage of foliage. Painting technology proposes the sputtering of products whose rheology may sometimes prove challenging.

Generally speaking, the production of sprays results from a flow instability.

The natural instability of a liquid cylinder, or the Rayleigh instability, is the prototype. Properly analyzed, it is rather a laboratory object, as the available flow rates are very small. It is, however, important to note that it is the only sputtering mode that generates single size drops.

The design of an atomizer is most often empirical. Generally speaking, an injector generates a sheet flow, a jet flow or another type of flow that tends to get destabilized. The flow may be naturally unstable (e.g. the conical sheet created by a household spray), or the instability may be promoted by the friction with a gaseous ambience or a gaseous flow.

A broad classification includes the mechanical injectors, which use the pressure of the liquid in a device whose geometry is apt to create an unstable flow, and aided injectors, in which the liquid flow is accompanied by a gaseous flow that tends to destabilize it.

1.3.3. *Spray characterization: particle sizing*

The drops that constitute a spray are never of the same size. Rayleigh instability is practically the only means to generate a one-dimensional spray. It is a laboratory device, quite delicate to implement in its controlled form, and which provides a very low flow rate.

In practice, the various sputtering processes produce polydisperse sprays.

They must be characterized by a statistical distribution of the size of drops.

Two types of information may be sought:

– the statistical law in detail;

– focus on mean diameters.

1.3.3.1. *Size distributions*

a) There are statistical distributions or probability function.

There are two main distributions: numerical and volumetric distributions.

The number fraction $f_n(x)$ is defined based on the number dn of drops whose diameter ranges between x and $x+\delta x$:

$$dn = f_n(x)\,\delta x \qquad [1.71]$$

The volume fraction $f_V(V)$ is defined based on the number dn_V of drops whose volume ranges between v and $v+\delta v$:

$$dn_V = f_V(v)\,\delta v \qquad [1.72]$$

By definition:

$$\int_0^\infty f_n(x)\,\delta x = 1 \qquad [1.73]$$

$$\int_0^\infty f_V(v)\,\delta v = 1 \qquad [1.74]$$

Cumulative laws are also defined.

For example, the volume cumulative $F_V(x)$ gives the fraction of liquid volume contained in the drops whose diameter exceeds the value x.

b) Mean diameters

Mean diameters can be defined using these distributions.

A mean diameter d_{pq} can thus be defined using the numerical distribution $f_n(x)$

$$d_{pq} = \int_0^{\infty} \left[\frac{x^p \ f_n(x)}{x^q \ f_n(x)} \right]^{\frac{1}{p-q}} dx \qquad [1.75]$$

The main mean diameters used are:

– the numerical mean diameter d_{10}

$$d_{10} = \int_0^{\infty} x f_n(x) \delta x \qquad [1.76]$$

– the volume mean diameter d_{30}

$$d_{30} = \left[\int_0^{\infty} x^3 \ f_n(x) \delta x \ \right]^{\frac{1}{3}} \qquad [1.77]$$

– the Sauter mean diameter d_{32}

$$d_{32} = \int_0^{\infty} \frac{x^3 \ f_n(x)}{x^2 \ f_n(x)} \delta x \qquad [1.78]$$

This diameter is very common in the literature, especially in the fields of combustion and engines. It implicitly contains a coupling between the surface and the volume. It will be seen further on that evaporation is a surface phenomenon, while combustion depends on the fuel richness throughout volumes.

c) Various statistical laws were used, the most common of which are:

– ***Rosin–Rammler law***

Initially used to characterize coal aggregates, it is probably the most well-known and at any rate the oldest law.

It relies on volume cumulative

$$F_V\left(d<x\right)=\exp-\left(\frac{x}{\overline{X}}\right)^N \qquad [1.79]$$

This law features two parameters $\overline{X}$ and N. These two parameters, one having the dimension of a diameter and the other one being dimensionless, are sufficient to provide all the information on the relative distribution of drop sizes. This law is all the narrower as N is larger.

A numerical density can be deduced from it:

$$f_n=\frac{N\,x^{n-4}}{\left(\overline{X}\right)^{N-3}\Gamma\left(1-\frac{3}{N}\right)}\exp\left(\frac{x}{\overline{X}}\right)^N \qquad [1.80]$$

with Gamma function:

$$\Gamma\left(x\right)=\int_0^\infty t^{x-1}\,e^{-t}dt \qquad [1.81]$$

– ***Log-normal law***

This law, proposed by Mugele and Avans, uses an auxiliary variable:

$$Y=Ln\,x-Ln\,\overline{X}=Ln\frac{x}{\overline{X}} \qquad [1.82]$$

It is a numerical law:

$$f_n\left(x\right)=\frac{N}{\sqrt{\pi}}\exp-N^2Y^2 \qquad [1.83]$$

Here, N is also an enlargement parameter.

Mean diameters are readily calculated:

$$d_{pq} = X \exp \frac{p+q-6}{4N^2} \qquad [1.84]$$

– ***Nukiyama–Tanasawa law***

This is a numerical density:

$$f_n = A\, x^2 \exp - \left(\frac{x}{\overline{X}}\right)^N \qquad [1.85]$$

For $N = 6$, it is similar to the Rosin–Rammler law.

Mean diameters can be calculated:

$$d_{pq} = \left(\overline{X}\right)^{p-q} \frac{\Gamma\left(\dfrac{p+3}{N}\right)}{\Gamma\left(\dfrac{q+3}{N}\right)} \qquad [1.86]$$

One variant is the ***Tanasawa–Tesima*** law:

$$f_n = A\, x^{\alpha} \exp - \left(\frac{x}{\overline{X}}\right)^{\beta} \qquad [1.87]$$

This law is similar to the **Weibull law**, which is common in the theory of reliability:

$$f_n = \left(\frac{x}{N}\right)^N \exp - \left(\frac{x}{\overline{X}}\right)^N \qquad [1.88]$$

d) Particle sizing has historically employed various metrology methods:

– The impaction method, derived from solid dust technologies.

– The photographic method: with large magnification with brief lighting (spark, then laser).

– Optical diffraction methods: the spray is dealt with similar to a pupillary function in a Fraunhofer setup. The difficulty is twofold: having an annular detector, which has been available for several decades now, and having a technique for the inversion of the diffraction figure, which was more challenging throughout history.

1.3.4. *Spray vaporization: the case of the isolated drop*

Mist vaporization is common in chemical engineering sciences and is systematically encountered in car or spaceship engines. Indeed, combustion is a phenomenon of gaseous phase and the introduction of a liquid fuel in a combustion chamber requires its preliminary vaporization. Calculating spray vaporization is therefore a must.

This type of calculation relies on the theory of vaporization of the isolated drop.

The Godsave model is a basis for this evaluation. It deals with vaporization controlled by the contribution of heat conduction or convection to the drop.

This model is not always sufficient. The problem is in fact complex. Vapor diffusion in the ambient atmosphere may oppose resistance to the transfer. Moreover, fuels that are commonly used in the automobile or aerospace fields are petroleum cuts, therefore mixtures of species. Other fuels are used in the aerospace industry: cryogenic fuels, sometimes in supercritical state, or propellants (containing the same oxidizing and fuel products).

To remain within the scope of this book, the focus here is on providing the results of Godsave's calculation.

This theory gives a law that is known as the D^2 -law.

The vaporization of the drop involves the decreases of its diameter in time. The model yields the following relation:

$$D^2(t) = D_0^2(t) - K\,t \qquad [1.89]$$

Constant K is calculated as follows:

$$K = 8\frac{\lambda_G}{c_{PG}\rho_L}Ln\left(1+B\right) \qquad [1.90]$$

where B, known as the Spalding parameter, is given by:

$$B = \frac{C_{PG}\left(T_V - T_a\right)}{L_V} \qquad [1.91]$$

The constant can be corrected in the presence of forced convection. Given Reynolds' values of drops in practice, laminar convection is assumed. The correction yields:

$$K_{conv} = K\left(1 + 0.3\,R_D^{0.5}\,\mathrm{Pr}^{1/3}\right) \qquad [1.92]$$

Whittaker proposed a more complex correction law, relying on a Nusselt number:

$$Nu_D = 2 + \left[0.4\,R_D^{0.5} + 0.06\,R_D^{2/3}\right] P_r^{0.4}\left(\frac{\mu_e}{\mu_w}\right) \qquad [1.93]$$

$Nu_D = 2$ then corresponds to pure conduction without convection.

This formula would be valid for $3.5 < R_e < 80\,000$ and $0.7 < P_r < 380$.

The correction would then be:

$$K_{conv} = K\left(1 + \left[0.2\,R_D^{0.5} + 0.03\,R_D^{2/3}\right] P_r^{0.4}\left(\frac{\mu_e}{\mu_w}\right)\right) \qquad [1.94]$$

1.4. Several examples

As already mentioned, the reader may lack the skills required to design a nuclear power plant or a refinery. In the absence of such skills, the reader would need to refer to a literature covering a wider range of knowledge. This includes in particular the calculation of complex two-phase transfers related to a specific numerical modeling (see [BER 81]).

Consequently, the following examples focus on the physics of two-phase systems and on sprays, domains that are somewhat ignored by certain textbooks.

EXAMPLE 1.1.– Determination of heat of vaporization

Using the tables provided by Appendix 3 and Clapeyron's formula, find the heat of vaporization of water under an atmospheric pressure of $1.013\,bar$ and at a temperature of $225\,C$. Compare it to the values given by the same tables.

SOLUTION TO EXAMPLE 1.1.–

The two chosen operating points correspond to two vaporization plateaus:

$$1\,bar = 10^5\ Pa = 0.1\,MPa$$

The plateau corresponding to $1\,bar$ is therefore characterized by

$$p = 1.013\,MPa$$

$$T = 100\,C$$

The plateau corresponding to $225\,C$ is characterized by

$$p = 2.548\,MPa$$

$$T = 225\,C$$

Recalling Clapeyron's formula:

$$\frac{d\,p_{Sat}}{d\,T} = \frac{L}{T\left(v_V - v_L\right)} \qquad [1.95]$$

Working on a table, the pressure derivative must be calculated using the "finite differences" method:

$$\frac{d\,p}{d\,T} \approx \frac{\Delta\,p}{\Delta\,T} \qquad [1.96]$$

The calculation can be performed between the value on the plateau and the lower or higher value of pressure.

For the first plateau:

$$T = 100\ C = 373.15\ K$$

$$v_V - v_L = 1.6729 - 0.001044 = 1.6718\ m^3\ kg^{-1} \qquad [1.97]$$

In "low" values:

$$\Delta p = 101.3 - 84.55 = 16.75\ kPa \qquad [1.98]$$

$$\Delta T = 100 - 95 = 5\ C \qquad [1.99]$$

In "high" values:

$$\Delta p = 120.82 - 101.3 = 19.52\ kPa \qquad [1.100]$$

$$\Delta T = 105 - 100 = 5\ C \qquad [1.101]$$

For the second plateau:

$$T = 225\ C = 498.15\ K$$

$$v_V - v_L = 0.07849 - 0.001199 = 0.07729\ m^3\ kg^{-1} \qquad [1.102]$$

In "low" values:

$$\Delta p = 2.548 - 2.318 = 0.23 MPa \qquad [1.103]$$

$$\Delta T = 225 - 220 = 5\ C \qquad [1.104]$$

In "high" values:

$$\Delta p = 2.795 - 2.548 = 0.247 MPa \qquad [1.105]$$

$$\Delta T = 230 - 225 = 5\ C \qquad [1.106]$$

In each of these four cases, the heat of vaporization can be calculated using the expression deduced from Clapeyron's formula:

$$L \approx T\left(v_V - v_L\right)\frac{\Delta p_{Sat}}{\Delta T} \qquad [1.107]$$

For the first plateau:

Low values:

$$L \approx T(v_V - v_L)\frac{\Delta p_{Sat}}{\Delta T} = \frac{373.15 * 1.6718 * 16.75.10^3}{5} = 2089\ kJ.kg^{-1} \quad [1.108]$$

High values:

$$L \approx T(v_V - v_L)\frac{\Delta p_{Sat}}{\Delta T} = \frac{373.15 * 1.6718 * 19.52.10^3}{5} = 2435\ kJ.kg^{-1} \quad [1.109]$$

For the second plateau:

Low values:

$$L \approx T(v_V - v_L)\frac{\Delta p_{Sat}}{\Delta T} = \frac{498.15 * 0.07729 * 230.10^3}{5} = 1771\ kJ.kg^{-1} \quad [1.110]$$

High values:

$$L \approx T(v_V - v_L)\frac{\Delta p_{Sat}}{\Delta T} = \frac{498.15 * 0.07729 * 247.10^3}{5} = 1902\ kJ.kg^{-1} \quad [1.111]$$

Let us compare these to the "actual" values.

The values $2089\ kJ.kg^{-1}$ and $2435\ kJ.kg^{-1}$ must be compared to $L = 2257\ \ kJ.kg^{-1}$. The result obtained is within 7 or 8%.

It is important to note that the average $\dfrac{2089 + 2435}{2} = 2262\,kJ.kg^{-1}$ gives a significantly better approximation.

$1771\ kJ.kg^{-1}$ and $1902\ kJ.kg^{-1}$ must be compared to $L = 1836{,}5\ \ kJ.kg^{-1}$. The result obtained is within 3.5%.

It is important to note that the average $\dfrac{1771 + 1902}{2} = 1836\,kJ.kg^{-1}$ gives a significantly better approximation.

NOTE.– Given the above notes, it is recommended to calculate the derivative of the pressure using values that frame the studied plateau:

$$\frac{\Delta p_{Sat}}{\Delta T} = \frac{(16.75+19.52)10^3}{10}, \text{ or } \frac{\Delta p_{Sat}}{\Delta T} = \frac{(230+247)10^3}{10} \quad [1.112]$$

EXAMPLE 1.2.– An intriguing bubble

An astute physicist managed to obtain a benzene bubble of diameter $D_1 = 2R_1 = 1cm$ within another water bubble of diameter $D_2 = 2R_2 = 2cm$. The pressure of the ambient air external to the two bubbles is $p_a = 1000\ Pa$. The two bubbles contain air. Within the entire system, there is a constant temperature equal to $T_a = 20C$.

What is the pressure inside the benzene bubble?

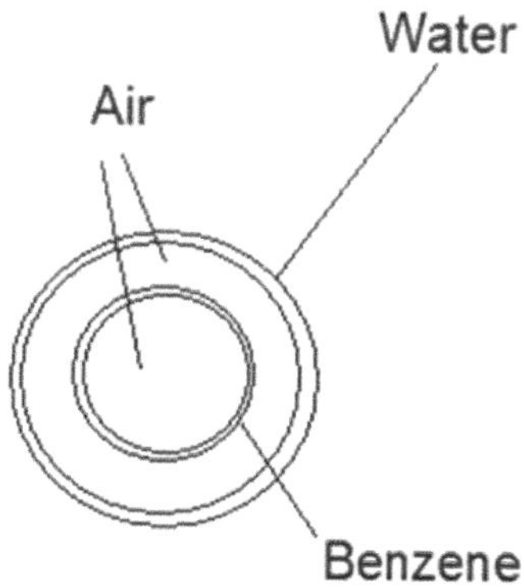

Figure 1.12. *Benzene bubble*

The values of surface tension for water and benzene, which are respectively $\sigma = 7.2.10^{-2}\ N\,m^{-1}$ and $\sigma = 0.44\ N\,m^{-1}$, are given.

SOLUTION TO EXAMPLE 1.2.–

Laplace's formula can be applied to each of the two bubbles. Since there are two interfaces to the bubble, the difference between the internal pressure and the external pressure is:

$$p_i - p_e = \frac{4\sigma}{R} \quad [1.113]$$

The pressure in the bubble of diameter D_2 is determined using $\sigma = 7.2.10^{-2}\ N\,m^{-1}$:

$$p_{i2} - p_a = \frac{4*7.2.10^{-2}}{10^{-2}} \quad [1.114]$$

Hence:

$$p_{i2} = 1000 + \frac{4*7.2.10^{-2}}{2.10^{-2}} 1014.4\,Pa \quad [1.115]$$

The pressure in the bubble of diameter D_1 is determined using $\sigma = 0.44\ N\,m^{-1}$:

$$p_{i1} - p_{i2} = \frac{4*0.44}{10^{-2}} \quad [1.116]$$

Hence:

$$p_{i2} = 1014.4 + \frac{4*0.44}{10^{-2}} = 1190.4\,Pa \quad [1.117]$$

EXAMPLE 1.3.– Identification by surface tension

A physicist is in the presence of a liquid. This can be a liquid A or a liquid B. Liquids A and B are indiscernible by their appearance, odor or any other first-hand characteristic.

The only data the physicist has on the liquids A and B are their density and their surface tension.

To discern whether the liquid is an A or a B, the physicist conducts the following experiment.

He pours in a vase a quantity of water, leaving a free surface. He then plunges in this water a transparent tube of unknown diameter. Water rises in the tube. The physicist measures the height of the water in the tube at about 10 mm above the free surface.

He replaces the water in the vase with his unknown liquid. The liquid then rises at about 4 mm above the free surface.

At the beginning and at the end of the experiment, the physicist checks the thermometer in the laboratory. It indicates $20\ C$.

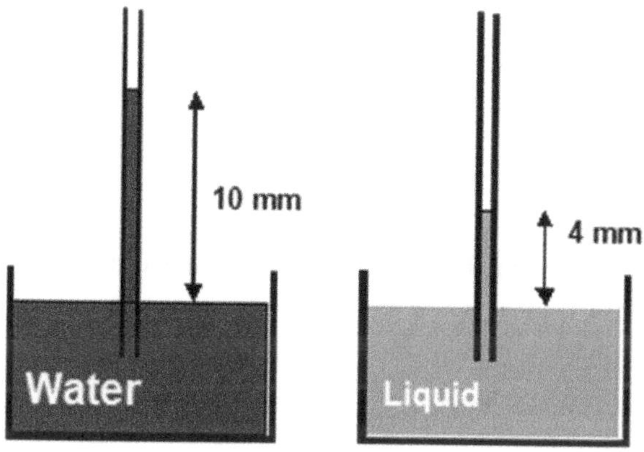

Figure 1.13. *Liquids*

1) Can the physicist state whether he is in the presence of a liquid A or B?

2) In fact, is it possible for him to know the diameter of the tube employed?

The properties of liquids A and B are given below.

Liquid	Surface tension $N.m^{-1}$	Density $kg.m^{-1}$
A	0.025	850
B	0.035	750

SOLUTION TO EXAMPLE 1.3.–

1) Jurin's law can be applied here:

$$h = \frac{4\sigma}{\rho g D} \qquad [1.118]$$

For liquids A and B, the physicist can write using the available data:

– for water:

$$0.1 = \frac{4\sigma_{water}}{10^3\, g\, D} \qquad [1.119]$$

– for the liquid:

$$0.04 = \frac{4\sigma}{\rho\, g\, D} \qquad [1.120]$$

By comparison, knowing that $\sigma_{water} = 7.2.10^{-2}$, they can write:

$$\frac{\sigma\,\rho_{water}}{\sigma_{water}\,\rho} = \frac{0.04}{0.1} \qquad [1.121]$$

or:

$$\frac{\sigma}{\rho} = 0.4 * \frac{7.2.10^{-2}}{10^3} = 2.88.10^{-5}\, N\, m^2\, kg^{-1} \qquad [1.122]$$

To be compared with:

– for liquid A:

$$\frac{\sigma}{\rho} = \frac{0.025}{850} = 2.94.10^{-5}\, N\, m^2\, kg^{-1} \qquad [1.123]$$

– for liquid B:

$$\frac{\sigma}{\rho} = \frac{0.035}{750} = 4.67.10^{-5}\, N\, m^2\, kg^{-1} \qquad [1.124]$$

Within the experimental precision (experiments involving surface tension are always very challenging), the physicist is in the presence of a liquid A.

2) Knowing $\frac{\sigma}{\rho}$, this value can be used in Jurin's law to determine the diameter of the tube; this can be done for water or liquid!

For water:

$$0.1 = \frac{4*7.2.10^{-3}}{100\,g\,D} \qquad [1.125]$$

or:

$$D = \frac{4*7.2.10^{-2}}{100*9.81*0.1} = 2.93.10^{-3}\,m = 2.93\,mm \qquad [1.126]$$

EXAMPLE 1.4.– Capillarity and the Eiffel Tower

A somewhat peculiar physicist wants to bring up water from ground level to the top of the Eiffel Tower, without pumping it.

Their setup for this purpose consists of immersing the bottom of a very fine vertical tube of height h, equal to that of the Eiffel Tower, into a water tank.

1) Knowing that starting with 2000 the height of the Eiffel Tower is evaluated at $h = 324\,m$, what should the inner diameter of this tube be?

2) Is the experiment feasible (financial aspects are not considered here)?

Figure 1.14. *Eiffel Tower*

SOLUTION TO EXAMPLE 1.4.–

1) This problem, which is rather a matter of "entertainment" than of technical study, is an application of Jurin's law:

$$h = \frac{4\sigma}{\rho g D} \qquad [1.127]$$

The standard value $\sigma = 7.2.10^{-2}\ N.m^{-1}$ is considered for the surface tension of water and $\rho = 1000\ kg.m^{-3}$ for the density.

Denoting by D the sought-after diameter, then:

$$D = \frac{4\sigma}{\rho g h} = \frac{4*7.2.10^{-2}}{1000*9.81*324} = 9.06.10^{-8}\ m = 9.06.10^{-2}\ \mu m \qquad [1.128]$$

Or still $90.6.10^{-2}\ nm$!

Manufacturing a capillary of $9.06.10^{-2}\ \mu m$ in diameter appears to be a very challenging task!

EXAMPLE 1.5.– Capillarity rise

A granita wall presents upward leaching up to $h = 40\ cm$ above the ground. What is the porosity of the material?

SOLUTION TO EXAMPLE 1.5.–

The porosity can be defined based on the diameter of the capillary d equivalent to the structure of the material.

Jurin's law can be used to readily calculate, employing the same choices of properties as in the previous problem

$$h = \frac{4\sigma}{\rho g d} \qquad [1.129]$$

or

$$d = \frac{4\sigma}{\rho g h} = \frac{4*7.2.10^{-2}}{1000*9.81*0.4} = 7.34.10^{-5}\ m = 73.4\ \mu m \qquad [1.130]$$

Unlike the previous problem, this is a concrete application!

EXAMPLE 1.6.– The sap rises

It has been assumed for a long time that the sap rises in trees by capillarity in narrow tubes present in the wood structure. Assimilating sap to water, what is the order of magnitude of diameter d of these narrow tubes that would allow the sap to rise from the ground to the highest branch of a tree whose height is 40 m?

SOLUTION TO EXAMPLE 1.6.–

Jurin's law can be used to write, knowing that $\sigma_{water} = 7.2.10^{-2}$ and $\rho = 10^3\ kg\, m^{-3}$

$$h = \frac{4\sigma}{\rho g d} \quad [1.131]$$

or:

$$40 = \frac{4 * 7.2.10^{-2}}{10^3\, g\, d} \quad [1.132]$$

Then:

$$d = \frac{4 * 7.2.10^{-2}}{10^3 * 9.81 * 40} = 7.34.10^{-7}\ m = 0.734\ \mu m \quad [1.133]$$

EXAMPLE 1.7.– Evaporation of a water spray

A water spray is very rapidly introduced in a chamber with air heated at $T_a = 400 C$.

The drops composing the spray are assumed to all have the same diameter $D_0 = 50\,\mu m$.

The latent heat of vaporization of water is given by

$$L_V = A - BT \quad [1.134]$$

where: $A = 3334\ kJ\, kg^{-1}$

$B = 2.9\ kJ\, kg^{-1} K^{-1}$

At $T = 400\ C$, the thermal conductibility of the air is $\lambda = 0.0509\ W\ m^{-1}\ K^{-1}$ and its specific heat capacity is $c_{PG} = 1004\ J\ Kg^{-1}$.

1) How long does it take for 45% of this spray to evaporate?

2) What is the full vaporization time of this spray?

SOLUTION TO EXAMPLE 1.7.–

1) According to Godsave's law without convection:

$$D^2(t) = D_0^2(t) - K\,t$$

Therefore, the constant K should be calculated:

$$K = 8\frac{\lambda_G}{c_{PG}\rho_L} Ln(1+B) = 8\frac{\lambda_G}{c_{PG}\rho_L} Ln\left(1+\frac{c_{PG}(T_V - T_a)}{L_V}\right) \quad [1.135]$$

Here, T_V is the saturation temperature of water, $T_V = 100C$ and $T_a = 400C$.

Moreover, the latent heat of vaporization of water is:

$$L_V = 3334 - 0.29 * 373 = 2252\ kJ\ kg^{-1}$$

Here, K is given by:

$$K = 8*\frac{0.0509}{1004*1000} Ln\left(1+\frac{1004*300}{2252.10^3}\right) \quad [1.136]$$

$$K = 4.95.10^{-8}\ m^2\ s^{-1}$$

The spray is 45% vaporized when the volume of each drop represents 55% of its initial value. Then, the diameter D of each drop is such that:

$$\left(\frac{D}{D_0}\right)^3 = 0.55 \quad [1.137]$$

which yields:

$$D = (0.55)^{1/3} * 50.10^{-6} = 41.10^{-6}\ m \quad [1.138]$$

This vaporization is obtained at τ_{V1}

$$\tau_{V1} = \frac{D_0^2 - D^2}{K} = \frac{2500.10^{-12} - 1681.10^{-12}}{4.95.10^{-8}} \quad [1.139]$$

$$\tau_{V1} = 0.0165\,s = 16.5\ ms$$

2) The total time of vaporization τ_{V2} is given by

$$\tau_{V2} = \frac{D_0^2}{K} = \frac{2500.10^{-12}}{4.95.10^{-8}} \quad [1.140]$$

$$\tau_{V2} = 0.05\,s = 50\,ms$$

EXAMPLE 1.8.– Evaporation of a sodium spray

A sodium spray evaporates in a nitrogen atmosphere at $T_a = 1500\,C$ under a pressure of $76\,cm$ Hg.

This spray has a simple histogram that can be summarized at $t = 0$ through a distribution between four sizes of drops.

For 100 drops of spray:

Diameter of the drop in micron	**Number within size for 100 drops**
10	**15**
50	**35**
100	**35**
200	**15**

1) What is the temperature at the surface of sodium drops under evaporation?

2) What is the vaporization constant K of the sodium drops?

3) What is the numerical mean diameter of this distribution, d_{10} at the moment $t = 0$?

4) Consider this spray after $\tau_V = 240\ ms$.

What is the new distribution for 100 drops? It is recommended to calculate the evolution of the drop sizes starting with the largest size class.

5) For this new distribution, what are at the moment $t = \tau_V$, the numerical mean diameter (d_{10}), volume mean diameter (d_{30}) and Sauter mean diameter or SMD (d_{32})?

Given:

Density of liquid sodium: $\rho_L = 927\ kg\ m^{-3}$

Enthalpy of vaporization of sodium: $L_V = 96.96\ kJ\ mol^{-1}$

Heat capacity of air: $c_{PG} = 1004\ J\ kg^{-1}$

Thermal conductibility of air: $\lambda_G = 0.024\ W\ m^{-1}\ K^{-1}$

Given: $Na = 23$

The melting temperature of sodium is $T_f = 371\ K$.

SOLUTION TO EXAMPLE 1.8.–

1) The surface of drops is at the melting temperature:

$$T_f = 371\ K = 98C$$

2) The evolution of the size of each droplet is given by the D^2 -law:

$$D^2(t) = D_0^2(t) - K\,t \qquad [1.141]$$

The sodium vaporization constant can be calculated as:

$$K = 8\frac{\lambda_G}{c_{PG}\rho_L}Ln(1+B) = 8\frac{\lambda_G}{c_{PG}\rho_L}Ln\left(1+\frac{c_{PG}(T_V - T_a)}{L_V}\right) \qquad [1.142]$$

The heat of vaporization can be deduced from the enthalpy of vaporization per mole, whose mass is $M = 23\,g = 23.10^{-3}\,kg$

$$L_V = \frac{96.96.10^3}{23.10^{-3}} = 4.21.10^6\,J\,kg^{-1} \qquad [1.143]$$

K is:

$$K = 8\frac{0,024}{1004*927}Ln\left(1+\frac{1004*(T_V - T_a)}{4.21.10^6}\right) = 5.94.10^{-8} \qquad [1.144]$$

3) According to the definition of mean diameters:

$$d_{pq} = \int_0^{\infty}\left[\frac{x^p\,f_n(x)}{x^q\,f_n(x)}\right]^{\frac{1}{p\_q}} dx \qquad [1.145]$$

then:

$$d_{10} = \int_0^{\infty} x\,f_n(x)\,\delta x \qquad [1.146]$$

The histogram over drops can be used to readily determine a table of the distribution function:

Drop diameter size in micron	**Number in the size** $f_n(x)$
10	**0.15**
50	**0.35**
100	**0.35**
200	**0.15**

Hence, the numerical mean diameter:

$$d_{10} = (0.15*10)+(0.35*50)+(0.35*100)+(0.15*200) = 84\,\mu m \qquad [1.147]$$

4) After $t = \tau_V = 240.10^{-3}\,s$, the size per class of drops can be calculated using $K = 5.94.10^{-8}$.

Initial size of $200\,\mu m$:

$$D^2 = \left(200.10^{-6}\right)^2 - 5.94.10^{-8} * 0.24 = 2.57.10^{-8} \quad [1.148]$$

$$D = 1.6.10^{-4}\ m = 160\,\mu m$$

Initial size of $100\,\mu m$:

$$D^2 = \left(100.10^{-6}\right)^2 - 5.94.10^{-8} * 0.24 = 4.26.10^{-9}$$

$$D = 6.53.10^{-5}\ m = 65.3\,\mu m$$

Initial size of $50\,\mu m$:

$$D^2 = \left(50.10^{-6}\right)^2 - 5.94.10^{-8} * 0.24 = -1.17.10^{-8}$$

This negative result makes no physical sense; in fact, the class $D = 50\,\mu m$ is fully vaporized. This can be verified by calculating the vaporization time:

$$t_{V50} = \frac{D_0^2}{K} = \frac{\left(50.10^{-6}\right)^2}{5.94.10^{-8}} = 4.21.10^{-2}\ s = 42.1\,ms \quad [1.149]$$

Obviously, class $D = 50\,\mu m$ will also be fully vaporized.

The histogram of the final spray is then:

Size micron	Number
65.3	**35**
160	**15**

This makes it possible to calculate the numerical distribution function:

Size micron	$f_n(x)$
65.3	**0.70**
160	**0.30**

Mean diameters can then be readily calculated:

– numerical mean diameter:

$$d_{10} = (0.7 * 65.3) + (0.3 * 160) = 92.1\,\mu m \qquad [1.150]$$

– volume mean diameter:

$$d_{10} = \left\{\left[0.7 * (65.3)^3\right] + \left[0.3 * (160)^3\right]\right\}^{\frac{1}{3}} = 112.5\,\mu m \qquad [1.151]$$

– Sauter mean diameter:

$$d_{32} = \frac{\left[0.7 * (65.3)^3\right] + \left[0.3 * (160)^3\right]}{\left[0.7 * (65.3)^2\right] + \left[0.3 * (160)^2\right]} = 68.1\,\mu m \qquad [1.152]$$

EXAMPLE 1.9.– Bubble dynamics

This problem requires a purely analytical resolution. There are no numerical values to be considered.

The evolution of a bubble in a liquid is considered. This bubble is *exclusively filled with vapor* of this liquid at the saturation vapor pressure.

The following are considered negligible:

– viscosity effects;

– surface tension effects.

The bubble is initially at equilibrium with a radius R_0. The ambient pressure is then $p_{\infty 0}$.

The ambient pressure follows the law:

$$p_\infty(t) = p_V - p_{\infty 0} \exp(2\alpha t) \qquad [1.153]$$

p_V, pressure of vapor at the temperature of the experiment is the pressure in the bubble.

1) Considering the previous hypotheses, what is the differential equation describing the temporal evolution $R(t)$ of the radius of the bubble?

2) It can be noted that the radius of the bubble $R(t)$ will have the form:

$$R(t) = A\exp(2\alpha t) \qquad [1.154]$$

where A is a constant. Determine A.

What can be said about the initial radius R_0?

SOLUTION TO EXAMPLE 1.9.–

1) The Rayleigh–Plesset equation describes the evolution of the radius of a bubble submitted to an ambience $p_\infty(t)$; without simplifying the hypotheses, it is written as:

$$R\frac{d^2R}{dt^2} + \frac{3}{2}\left(\frac{dR}{dt}\right)^2 + \frac{4\nu_L}{R}\frac{dR}{dt} + \frac{2\sigma}{\rho_L R} + \frac{\Delta p(t)}{\rho_L} = 0 \qquad [1.155]$$

Neglecting the effects of viscosity and surface tension, it is written as:

$$R\frac{d^2R}{dt^2} + \frac{3}{2}\left(\frac{dR}{dt}\right)^2 + \frac{p_\infty(t) - p_V}{\rho_L} = 0 \qquad [1.156]$$

with the boundary conditions:

$$t = 0 \quad ; \quad R = R_0$$

2) Intuitively, the solution to be found has the form:

$$R = A\exp\beta t$$

Noting that:

$$p_\infty(t) - p_V = -\,p_{\infty 0}\exp 2\alpha t \qquad [1.157]$$

and:

$$\frac{dR}{dt} = A\beta \exp \beta t \qquad [1.158]$$

$$\frac{d^2 R}{dt^2} = A\beta^2 \exp \beta t \qquad [1.159]$$

the simplified Rayleigh–Plesset equation is written as:

$$A e^{\beta t}\, A\beta^2 e^{\beta t} + \frac{3}{2}\left(A\beta\, e^{\beta t}\right)^2 = \frac{p_{\infty 0}\, e^{2\alpha t}}{\rho_L} \qquad [1.160]$$

Let:

$$A^2\beta^2 e^{2\beta t} + \frac{3}{2} A^2\beta^2 e^{2\beta t 2} = \frac{p_{\infty 0}\, e^{2\alpha t}}{\rho_L} \qquad [1.161]$$

and by identifying:

$$\alpha = \beta$$

$$\frac{5}{2} A^2\beta^2 = \frac{5}{2} A^2\alpha^2 = \frac{p_{\infty 0}}{\rho_L} \qquad [1.162]$$

$$\frac{5}{2} A^2\beta^2 = \frac{5}{2} A^2\alpha^2 = \sqrt{\frac{2p_{\infty 0}}{5\rho_L \alpha^2}} \qquad [1.163]$$

Hence:

$$R(t) = \sqrt{\frac{2p_{\infty 0}}{5\rho_L \alpha^2}} \exp 2\alpha t \qquad [1.164]$$

The value of the initial radius is:

$$R_0 = \sqrt{\frac{2p_{\infty 0}}{5\rho_L \alpha^2}} \qquad [1.165]$$

Therefore, R_0 and α cannot be independent.

2

Convection of Species

2.1. Introduction

Mass transfer is a major problem in the field of chemical engineering, where it is often coupled with heat transfer.

This association is the basis of a new ***specialized field, namely aerothermochemistry***. Besides chemical engineering, this field integrates combustion, which cannot be reduced to a complex chemistry, as it strongly couples reactions to heat and mass transports. A simple flame, such as that of a candle, closely associates gaseous flow and the reaction system.

Although the theme of mass convection is beyond the scope of this series of three volumes dedicated to heat transfer, the similarity that often appears in the equations modeling the phenomena justifies this short chapter, which is intended as a note on transfer similarities.

The mass transfer equations are established and developed in Appendices 4 and 5 of Volume 3.

2.2. Characterization of a mixture: definitions

The focus is here is on flows of fluids that are mixtures.

Consider a fluid of density ρ, mixture of n species, indexed by i.

The literature contains a variety of methods for defining the local composition of a mixture. Various definitions for the concentration of a species in a mixture can be used:

– ***molar concentration***: number of moles of the species i per unit volume in $mol.m^{-3}$;

– ***mole fraction***: ratio of the number of moles of species i to the total number of moles, dimensionless;

– ***mass concentration*** m_i: mass of the species per unit volume in $kg.m^{-3}$;

– ***mass fraction*** c_i: ratio of the mass of species i to the total mass per unit volume ρ, dimensionless $c_i = \dfrac{m_i}{\rho}$.

For pure chemistry purposes, molar concentration may be preferred. In the context of this chapter, mass fraction appears better adapted for expressions in the mechanics of fluids.

The following relation is obvious:

$$\sum_i c_i = \sum_i \frac{m_i}{\rho} = \frac{\sum_i m_i}{\rho} = \frac{\rho}{\rho} = 1 \qquad [2.1]$$

2.3. Law of diffusive transfer: Fick's law

This section introduces the phenomenon that is in fact the basis for the analogies between heat transfer and mass transfer.

In a mixture of n species indexed by i, the distribution of mass fractions in the mixture volume may be homogeneous or heterogeneous. In the first case, in the absence of chemical reactions, the distribution of mass fractions does not evolve in time.

However, potentially existing concentration gradients may influence the mixture due to chemical reactions or convective transports. A diffusion of species takes place.

A heterogeneous mixture is thus influenced by fluxes of diffusive mass.

This flux is defined by considering the mass of a species i that crosses a given surface *S*.

If this surface is arbitrary, then a mass flux or a mass flow rate of the species appears, which is denoted by $\dot{m}$ and expressed in $kg\ s^{-1}$.

Molar fluxes can also be defined.

If this is a unit surface, then a mass flux density or a unit mass flow rate of the species is defined, which is denoted by J_i and expressed in $kg\ m^{-2} s^{-1}$.

Molar flux densities can also be defined.

Fick's law describes the phenomenon referred to as molecular diffusion. This is presented here in terms of mass fraction.

This law gives the mass flow rate $d\dot{m}_i$ of the species i in a mixture through a surface dS of normal $\vec{n}$:

$$d\dot{m}_i = -\rho D_i \, \vec{grad}\, c_i \, \vec{n} \, dS \qquad [2.2]$$

which amounts to giving a diffusion mass flux density:

$$J_i = -\rho D_i \, \vec{grad}\, c_i \, \vec{n} \qquad [2.3]$$

The "$-$" sign means that the flow rate goes in the direction of descending concentrations.

Here, D_i is the diffusion coefficient of species c_i in the mixture, which is expressed in ms^{-2}. For a Fick's law expressed in molar concentrations, this coefficient will have an obviously different value.

An obvious analogy with the Fourier law can be noted here:

$$\varphi = -\lambda \, \overrightarrow{grad}\, T \qquad [2.4]$$

This analogy is not purely formal. The mechanism that determines the heat transport is the same as the one that induces molecular diffusion: it is thermal agitation.

Indeed, according to Boltzmann, similar to thermal conduction, diffusion is due to the effect of molecular collisions in a fluid. It will be seen that, in many simple cases (e.g. two-dimensional boundary layer), this leads to a similarity of the equations of heat and mass transport in a fluid.

It is also important to note that the diffusion coefficient D_i defined here has the same dimension as the kinematic viscosity ν and the thermal diffusivity a, which are all expressed in ms^{-2}. Their respective roles are similar in the equations of impulse, energy and continuity of species.

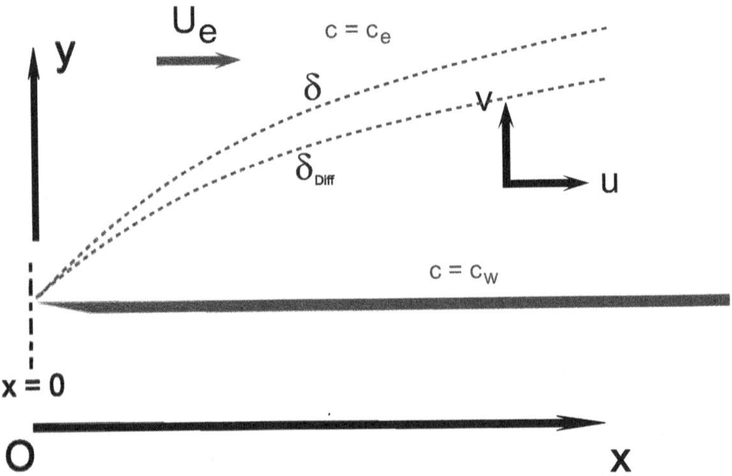

Figure 2.1. *Schematic representation of a diffusion boundary layer. For a color version of this figure, see www.iste.co.uk/ledoux/heat4.zip*

2.4. Similarity of transport equations

As this book focuses on convection, this section relies on the similarities of the equations of the non-reactive boundary layer, along with common simplifying hypotheses. Here, the diffusion layer is inert (non-reactive).

We give the classical equations:

$$\frac{\partial u}{\partial x}+\frac{\partial v}{\partial y}=0 \quad [2.5]$$

$$u\frac{\partial u}{\partial x}+v\frac{\partial v}{\partial y} = -\frac{1}{\rho}\frac{d\,p_e}{d\,x}+\nu\frac{\partial^2 u}{\partial y^2} \qquad [2.6]$$

$$u\frac{\partial T}{\partial x}+v\frac{\partial T}{\partial y} = a\frac{\partial^2 T}{\partial y^2} \qquad [2.7]$$

$$u\frac{\partial c_i}{\partial x}+v\frac{\partial c_i}{\partial y} = D_i\frac{\partial^2 c_i}{\partial y^2} \qquad [2.8]$$

NOTE.– The last line represents n independent equations.

The general continuity equation is not an $(n+1)th$ equation.

Indeed, summing up the n equations for each i gives:

$$\sum_i \rho\frac{d\,c_i}{d\,t}=\sum_i -div\,\rho\,c_i\,\vec{V}+\sum_i Q_i \qquad [2.9]$$

Mass conservation yields the relation:

$$\sum_i Q_i = 0 \qquad [2.10]$$

$$\sum_i \rho\frac{d\,c_i}{d\,t}=\sum_i -div\,\rho\,c_i\,\vec{V} \qquad [2.11]$$

$$\sum_i \frac{d\,\rho c_i}{d\,t}-\sum_i c_i\frac{d\,\rho}{d\,t}=\sum_i -div\,\rho\,c_i\,\vec{V} \qquad [2.12]$$

The continuity equation is thus found.

As already noted, kinematic viscosity and thermal diffusivity have symmetric roles.

The symmetry of the roles of kinematic viscosity and the diffusion coefficient, as well as that of thermal diffusivity and the diffusion coefficient, should be noted.

These three coefficients have the same dimension M^2T^{-1}, which are expressed in m^2s^{-1}.

Their ratios are numbers that are very important, particularly in the theory of boundary layers and heat and mass transfer:

Hence, by analogy with the Prandtl number:

$$\Pr = \frac{\nu}{a} \qquad [2.13]$$

The Schmidt number of species i can be defined as:

$$Sc_i = \frac{D_i}{a} \qquad [2.14]$$

Moreover, by analogy with the Reynolds number:

$$R_L = \frac{V\,L}{\nu} \qquad [2.15]$$

And the thermal Péclet number:

$$Pe = \frac{L\,V}{a} \qquad [2.16]$$

the Péclet number of species i can be found:

$$Pe_i = \frac{L\,V}{D_i} \qquad [2.17]$$

Finally, chemical reactions between species are possible. They are defined by the relations between c_i. Since they widely differ from one problem to another, they are not described here.

As a result of reactions, appearances or disappearances are possible, which are respectively associated with (positive) terms of creation or (negative) terms of disappearance of species i, or mass terms $Q_i\left(c_j\right)$. These terms expressed in $kg.m^{-3}$ depend a priori on all or part of c_j and on the kinetic constants of the reaction system specific to the problem being studied.

Moreover, according to chemistry:

$$\sum_i Q_i = 0 \tag{2.18}$$

A general expression of the mass transport can be found in the equation of transport of a species i, as in equation [2.8].

2.5. Solving the equations of diffusive transport

2.5.1. *Analogies*

Diffusive transport may be independent of thermal fields. Consider the case when the boundary layer is non-reactive (boundary layer of pure diffusion).

In this case, $Q_i \equiv 0$ [2.19]

$$Nu_L = 0.036\, R_L^{0.8}\, \mathrm{Pr}^{1/3} \tag{2.20}$$

Then, the system to be solved is the following:

$$\frac{\partial u}{\partial x} + \frac{\partial v}{\partial y} = 0 \tag{2.21}$$

$$u\frac{\partial u}{\partial x} + v\frac{\partial v}{\partial y} = -\frac{1}{\rho}\frac{d\,p_e}{d\,x} + \nu\frac{\partial^2 u}{\partial y^2} \tag{2.22}$$

$$u\frac{\partial c_i}{\partial x} + v\frac{\partial c_i}{\partial y} = D_i\frac{\partial^2 c_i}{\partial y^2} \tag{2.23}$$

which is the equivalent of the system found in the case of a purely thermal problem:

$$\frac{\partial u}{\partial x} + \frac{\partial v}{\partial y} = 0 \tag{2.24}$$

$$u\frac{\partial u}{\partial x} + v\frac{\partial v}{\partial y} = -\frac{1}{\rho}\frac{d\,p_e}{d\,x} + \nu\frac{\partial^2 u}{\partial y^2} \tag{2.25}$$

$$u\frac{\partial T}{\partial x} + v\frac{\partial T}{\partial y} = a\frac{\partial^2 T}{\partial y^2} \tag{2.26}$$

As will be noted, there may be differences between the problems, according to the two natures of transfer. Therefore various sets of boundary conditions should be considered.

In the field of mass transport, wall phenomena (catalysis, porous walls, etc.) may lead to the creation or disappearance of species comparable to mass fluxes at the wall.

The signaled analogies may be extended by the definition of dimensionless criteria related to this wall mass transfer.

The Nusselt number

$$Nu_L = \frac{hL}{\lambda} = \frac{\varphi_W L}{\lambda(T_e - T_W)} \qquad [2.27]$$

has an equivalent based on the mass flux density of species i; this is the Sherwood number:

$$Sh_{ix} = \frac{J_W x}{\rho D_i (c_{ie} - c_{iW})} \qquad [2.28]$$

Similarly, the Stanton number

$$St_x = \frac{\varphi_W}{\rho c_P U_e (T_e - T_W)} \qquad [2.29]$$

has an equivalent, namely the diffusive Stanton number:

$$St_{Diffx} = \frac{\varphi_W x}{\rho D U_e (c_{ie} - c_{iW})} \qquad [2.30]$$

Moreover, the relation between the Stanton and Nusselt numbers can be found between the diffusive Stanton number and the Sherwood number:

$$St = Nu_D \, R_D^{-1} \, \mathrm{Pr}^{-1} \qquad [2.31]$$

$$St_{Diffx} = \frac{J_W}{\rho U_e (c_{ie} - c_{iW})} = \frac{J_W x}{\rho D_i (c_{ie} - c_{iW})} \frac{\nu}{U_e x} \frac{D_i}{\nu} \qquad [2.32]$$

or

$$St_{Diffx} = Sh_{ix}\ R_x^{-1}\ Sc^{-1} \qquad [2.33]$$

2.5.2. *Non-reactive boundary layer: concentration fixed at the wall*

It may be that mass fractions of species i are specified at the wall and known in the external flow.

In this particular case, the solutions of the thermal transfer are formally transferrable to the diffusive transfer.

2.5.2.1. *Laminar boundary layer*

The following system should be solved:

$$\frac{\partial u}{\partial x} + \frac{\partial v}{\partial y} = 0 \qquad [2.34]$$

$$u\frac{\partial u}{\partial x} + v\frac{\partial v}{\partial y} = \nu\frac{\partial^2 u}{\partial y^2} \qquad [2.35]$$

$$u\frac{\partial c_i}{\partial x} + v\frac{\partial c_i}{\partial y} = D_i\frac{\partial^2 c_i}{\partial y^2} \qquad [2.36]$$

with the following boundary conditions:

$$y = 0\,;\, c_i = c_{iW} = Ct \qquad [2.37]$$

$$y \to \infty\,;\, c_i \to c_{ie} \qquad [2.38]$$

which is the equivalent of the system found in the case of a purely thermal problem:

$$\frac{\partial u}{\partial x} + \frac{\partial v}{\partial y} = 0 \qquad [2.39]$$

$$u\frac{\partial u}{\partial x} + v\frac{\partial v}{\partial y} = \nu\frac{\partial^2 u}{\partial y^2} \qquad [2.40]$$

$$u\frac{\partial T}{\partial x} + v\frac{\partial T}{\partial y} = a\frac{\partial^2 T}{\partial y^2} \qquad [2.41]$$

with the boundary conditions:

$$y = 0\,;\, T = T_W = Ct \qquad [2.42]$$

$$y \to \infty\,;\, T \to T_e = Ct \qquad [2.43]$$

In this case, the equivalent of Nusselt and Stanton numbers will have expressions that are equivalent to the results of the laminar boundary layer at constant wall temperature. The Schmidt number then replaces the Prandtl number.

To the formulas giving the Nusselt number:

$$Nu_x = 0.332\ R_x^{0.5}\ \mathrm{Pr}^{1/3} \qquad [2.44]$$

$$Nu_L = 0.664\ R_L^{0.5}\ \mathrm{Pr}^{1/3} \qquad [2.45]$$

correspond the local Sherwood number and the mean Sherwood number over a length L of the sheet:

$$Sh_x = 0.332\ R_x^{0.5}\ Sc^{1/3} \qquad [2.46]$$

$$Sh_L = 0.664\ R_L^{0.5}\ Sc^{1/3} \qquad [2.47]$$

Similarly, to the thermal Stanton numbers:

$$St_x = 0.332\ R_x^{-0.5}\ \mathrm{Pr}^{-2/3}\ \frac{\partial^2 \Omega}{\partial u^2} \qquad [2.48]$$

$$St_L = 0.664\ R_L^{-0.5}\ \mathrm{Pr}^{-2/3} \qquad [2.49]$$

correspond the diffusive Stanton numbers:

$$St_{Diffx} = 0.332\ R_x^{-0.5}\ Sc^{-2/3} \qquad [2.50]$$

$$St_{DiffL} = 0.664\ R_x^{-0.5}\ Sc^{-2/3} \qquad [2.51]$$

Moreover, it is important to note the relation between the Stanton number and the friction coefficient:

$$C_{fx} = \frac{0.664}{\sqrt{R_x}} = 0.664\, R_x^{-0,5} \qquad [2.52]$$

$$St_x = \frac{C_{fx}}{2}\, Sc^{-2/3} \qquad [2.53]$$

2.5.2.2. *Turbulent boundary layer*

Using fairly parallel reasoning, the results obtained in the turbulent regime are also transposable.

The following relations are obtained:

$$Sh_x = 0.0288\, R_x^{0.8} Sc^{1/3} \qquad [2.54]$$

$$Sh_L = 0.036\, R_L^{0.8}\, \mathrm{Pr}^{1/3} \qquad [2.55]$$

$$St_{Diffx} = 0.0288\, R_x^{-0.2}\, Sc^{-2/3} \qquad [2.56]$$

$$St_{DiffL} = 0.036\, R_L^{-0.2}\, Sc^{-2/3} \qquad [2.57]$$

And similar to the laminar regime:

$$St_{Diffx} = \frac{C_{fx}}{2}\, P_r^{-2/3} \qquad [2.58]$$

2.5.3. *Non-reactive boundary layer: wall catalysis*

2.5.3.1. *In this type of problem, the core of the boundary layer is inert (non-reactive), therefore* $Q_i \equiv 0$.

Wall catalysis introduces a new type of boundary conditions. Excluding the thermal influence, the following system can be written as:

$$\frac{\partial u}{\partial x} + \frac{\partial v}{\partial y} = 0 \qquad [2.59]$$

$$u\frac{\partial u}{\partial x}+v\frac{\partial v}{\partial y} = \nu\frac{\partial^2 u}{\partial y^2} \qquad [2.60]$$

$$u\frac{\partial c_i}{\partial x}+v\frac{\partial c_i}{\partial y} = D_i\frac{\partial^2 c_i}{\partial y^2} \qquad [2.61]$$

with the boundary conditions:

$$y=0\,;\, J_{Wi}=\pm k_i\, c_{iW}^n \qquad [2.62]$$

$$y\rightarrow\infty\,;\, c_i\rightarrow c_{ie} \qquad [2.63]$$

The wall kinetics leads to the appearance or disappearance of the species at the wall. It can then be assimilated to a wall flux density J_{Wi}: the $\pm$ sign corresponds to an effect of creation or suppression of the species.

Here, k_i is a reaction constant, whose order is n; the most common values are $n=1$ and $n=2$.

The value n corresponds to a specified wall flux density. Attention should be paid to the fact that the note made in the thermal case is still valid. The use of formulas established in **section 2.4.2** is dangerous in the case of a constant flux (errors reaching or even exceeding 40%).

NOTE.– This case is restrictive. The kinetics may be of a higher order depending on several concentrations.

2.5.3.2. *Thermal analogy of first order*

The inert boundary layer with a first-order catalytic reaction is the equivalent of a specific thermal problem.

Let us consider the thermal boundary layer above a flat plate of thickness e composed of an insulating material of thermal conductibility λ_W. Let us impose a temperature T_{ref} below this plate (on the other side with respect to the fluid).

The temperature of the wall perceived by the fluid is no longer constant, but depends on the local flux density. Therefore, at the wall:

$$T_W(x)=T_{ref}-\frac{\varphi_W}{\lambda_W} \qquad [2.64]$$

Defining a reduced variable:

$$\theta = \frac{T_{ref} - T}{T_{ref} - T_e} \qquad [2.65]$$

leads to a system of equations:

$$\frac{\partial u}{\partial x} + \frac{\partial v}{\partial y} = 0 \qquad [2.66]$$

$$u\frac{\partial u}{\partial x} + v\frac{\partial v}{\partial y} = \nu\frac{\partial^2 u}{\partial y^2} \qquad [2.67]$$

$$u\frac{\partial \theta}{\partial x} + v\frac{\partial \theta}{\partial y} = a\frac{\partial^2 \theta}{\partial y^2} \qquad [2.68]$$

with the boundary conditions:

$$\varphi_W = \lambda_W \left(T_{ref} - T_e\right)\theta_W \qquad [2.69]$$

$$y \to \infty\,;\theta \to 1_e = Ct \qquad [2.70]$$

which is equivalent to what follows, considering:

$$C_i = \frac{c_i}{c_e} \qquad [2.71]$$

$$\frac{\partial u}{\partial x} + \frac{\partial v}{\partial y} = 0 \qquad [2.72]$$

$$u\frac{\partial u}{\partial x} + v\frac{\partial v}{\partial y} = \nu\frac{\partial^2 u}{\partial y^2} \qquad [2.73]$$

$$u\frac{\partial C_i}{\partial x} + v\frac{\partial C_i}{\partial y} = D_i\frac{\partial^2 C_i}{\partial y^2} \qquad [2.74]$$

with the boundary conditions:

$$y = 0 \, ; J_{Wi} = \pm k_i \; C_{iW} \qquad [2.75]$$

$$y \to \infty \, ; C_i \to 1 \qquad [2.76]$$

Using the analogies:

C_i with θ

$$J_{Wi} = \pm k_i \, C_{iW} \text{ with } \varphi_W = \lambda_W \left(T_{ref} - T_e \right) \theta_W \qquad [2.77]$$

J_{Wi} with φ_W

$$k_i \, c_e \text{ with } \lambda_W \left(T_{ref} - T_e \right) \qquad [2.78]$$

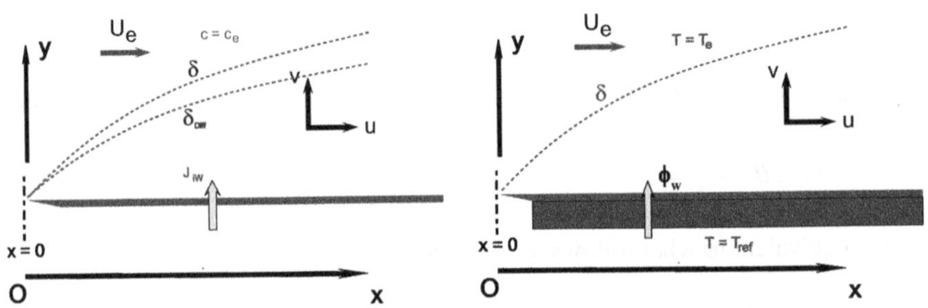

Catalysis: first-order wall reaction: $J_{Wi} = \pm k_i C_{iW}$ [2.79]

Thermal analogy

Figure 2.2. *Thermal analogy of the first-order wall reaction. For a color version of this figure, see www.iste.co.uk/ledoux/heat4.zip*

2.5.3.3. *Law of superposition*

The proposal was made [LED 75] to apply Lighthill's method of superposition to this analogy problem, which was presented in **Volume 3, Chapter 3**. It is important to recall that in the presence of a thermal transfer step (transfer starting at a distance x_0 from the leading edge), the classical formulas can be corrected by a function of $g\left(\dfrac{x}{x_0}\right)$.

Then:

$$St_x(x,x_0)=g\left(\frac{x}{x_0}\right)St_x(x,x_0=0) \qquad [2.80]$$

where $g\left(\frac{x}{x_0}\right)$ depends on the flow regime and the author proposing the formula.

In the laminar regime, the function $g\left(\frac{x}{x_0}\right)$ proposed by Eckert is:

$$g\left(\frac{x}{x_0}\right)=\frac{1}{\left[1-\left(\frac{x}{x_0}\right)^{3/4}\right]^{1/3}} \qquad [2.81]$$

And in the turbulent regime, the following relation proposed by Rubesin is:

$$g\left(\frac{x}{x_0}\right)=\frac{1}{\left[1-\left(\frac{x}{x_0}\right)^{39/40}\right]^{7/39}} \qquad [2.82]$$

or by Seban is:

$$g\left(\frac{x}{x_0}\right)=\frac{1}{\left[1-\left(\frac{x}{x_0}\right)^{9/10}\right]^{1/9}} \qquad [2.83]$$

These expressions can be written as:

$$St_x=g\left(\frac{x}{x_0}\right)St_{x,x_0=0} \qquad [2.84]$$

Here, x_0 is the abscissa for the beginning of the transfer and $St_{x,x_0=0}$ is the transfer without step.

Lighthill proposed a method [LIG 50] to determine the heat transfer for a specified distribution of wall temperature $T_W(x)$:

$$St_x = \int_0^x g\left(\frac{x}{\xi}\right) St_x(x, x_0 = 0)\, dT_W(\xi) \qquad [2.85]$$

This method can be adapted to the calculation of the wall flux of species i, in the presence of a catalysis step.

The mass flux of species i can by analogy be calculated at a point x, for a specified distribution of wall concentrations $c_i = c_i(x)$:

$$St_{Diff} = \int_0^x g\left(\frac{x}{\xi}\right) St_x(x, x_0 = 0)\, dT_W(\xi) \qquad [2.86]$$

This can be expressed in terms of flux, by deducing from the laminar or turbulent Stanton number the value of a mass convection coefficient:

$$St_{Diff} = \frac{J_{iW}}{\rho U_e(c_{iW} - c_{ie})} = \frac{h_i}{\rho U_e} \qquad [2.87]$$

Therefore:

$$J_{iW} = \int_0^x g\left(\frac{x}{\xi}\right) h(x, x_0 = 0)\, d\,c_{iW}(\xi) \qquad [2.88]$$

where $d\,c_{iW}(\xi)$ is the differential of the wall distribution law of the concentration of species i.

In the case of a wall catalysis, it can then be written that the mass flux density at the wall is linked to the concentration by a kinetic law:

$$J_{iW} = kc_{iW}^n. \qquad [2.89]$$

An integral equation is then obtained, where the function $c_{iW}(x)$ becomes unknown:

$$k\,c_{iW}^n = \int_0^x g\left(\frac{x}{\xi}\right) h(x, x_0 = 0)\, d\,c_{iW}(\xi) \qquad [2.90]$$

This equation will be solved numerically.

2.5.4. *Reactive boundary layer*

In the more general case of a reactive boundary layer, the system to be considered is more complex:

$$\frac{\partial u}{\partial x}+\frac{\partial v}{\partial y}=0 \quad [2.91]$$

$$u\frac{\partial u}{\partial x}+v\frac{\partial v}{\partial y} = -\frac{1}{\rho}\frac{d\,p_e}{d\,x}+\nu\frac{\partial^2 u}{\partial y^2} \quad [2.92]$$

$$u\frac{\partial T}{\partial x}+v\frac{\partial T}{\partial y} = a\frac{\partial^2 T}{\partial y^2}+Q_{react} \quad [2.93]$$

$$u\frac{\partial c_i}{\partial x}+v\frac{\partial c_i}{\partial y} = D_i\frac{\partial^2 c_i}{\partial y^2}+Q_i\left(c_j,T\right) \quad [2.94]$$

The equations can be strongly coupled. Indeed, the reactions can be endothermal or exothermal, hence the introduction of a term Q_{react} in the equation of energy.

Moreover, the reaction kinetics may depend on the temperature and other concentrations in the mixture, hence $Q_i=Q_i\left(c_j,T\right)$. Moreover, the number of species is highly variable from one problem to another. This case is common in chemical engineering, and it is a rule in combustion.

For this type of problem, the analogy does not work. There are no general solutions. The numerical resolution proves to be the best adapted approach.

3

Phase Change Materials

3.1. Introduction

The increase in power density of electronic components requires a more efficient heat management. Heat engineering is essential for the dimensioning of these components. Heat sinks (HS) based on phase change materials (PCMs) are widely used in several industrial applications: civil engineering, mechatronics, electrical engineering, etc. In particular, PCMs are mainly used in the field of electronic civil engineering [DEB 21]. Indeed, their thermal characteristics are ideal for an application such as building construction, allowing the best in situ management of heat transfer between inside and outside.

PCMs are materials able to modify their physical state for a determined and limited range of temperature.

When the material passes from the solid state to the liquid state, it absorbs heat, and when the reverse phenomenon occurs (temperature decrease), the material releases the previously absorbed heat, allowing heat transfer control.

The dominant phase change is melting/solidification and occurs within a quite large temperature range: [10–80]°C.

Consequently, the choice of the PCM will be determined by the temperature of the expected state change.

PCMs induce two possible types of heat transfer:

– **Thermal transfer by latent heat**:

The material stores or releases energy by phase change, maintaining a constant temperature (temperature characteristic for the phase change).

Indeed, the principle is to heat the material until it reaches its phase change temperature (solid to liquid, or liquid to gas). When it reaches this temperature, it absorbs an amount of heat required for the phase transformation (melting or vaporization latent heat).

If the reverse phenomenon occurs (from gas into liquid or liquid into solid), this amount of heat stored depends on the mass of the material.

– **Heat transfer by sensible heat**:

In this case, the material may release or store energy and its temperature may vary without a change of physical state. This type of heat storage is the most commonly used, and it is non-isothermal. It accumulates an amount of heat that is proportional to the mass of material and to the heat capacity.

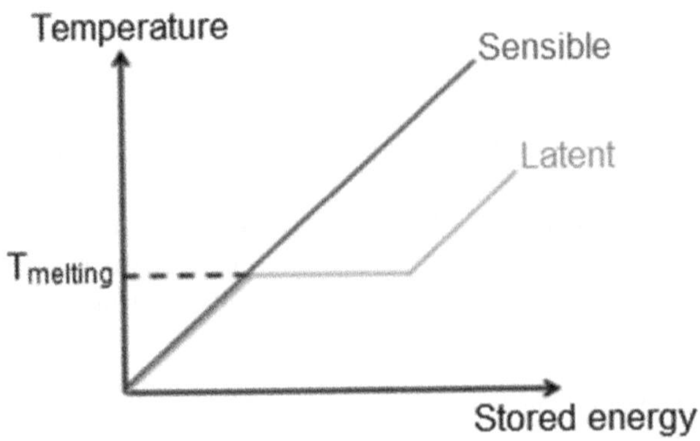

Figure 3.1. *Comparative representation of sensible and latent heat transfers. For a color version of this figure, see www.iste.co.uk/ledoux/heat4.zip*

There are many advantages to using PCMs.

Indeed, certain studies have demonstrated that in the building industry, PCMs are indestructible, inert, non-toxic and their maintenance requirements are low.

For example, Micronal PCM is used for buildings and experience has shown that 3 cm of coating containing only 30% of PCM was equivalent to 18 cm of concrete or 23 cm of brick.

This shows that these types of materials are of interest for small size applications such as electronic boards, thus allowing their net cooling for a small amount of PCM used.

There are many criteria for the selection of a PCM. They are at the same time kinetic, chemical, thermo-physical, and also ecological and economic.

One of the important characteristics in the choice of a PCM is its melting temperature, which must be consistent with the conditions and the range of phase change temperature (the shortest possible). Another important property is that the material has good thermal conductivity in order to facilitate and optimize heat transfer.

In order to obtain the highest possible density storage, PCMs must have the best volume phase change enthalpy.

Figure 3.2 groups the various types of PCMs and their classification:

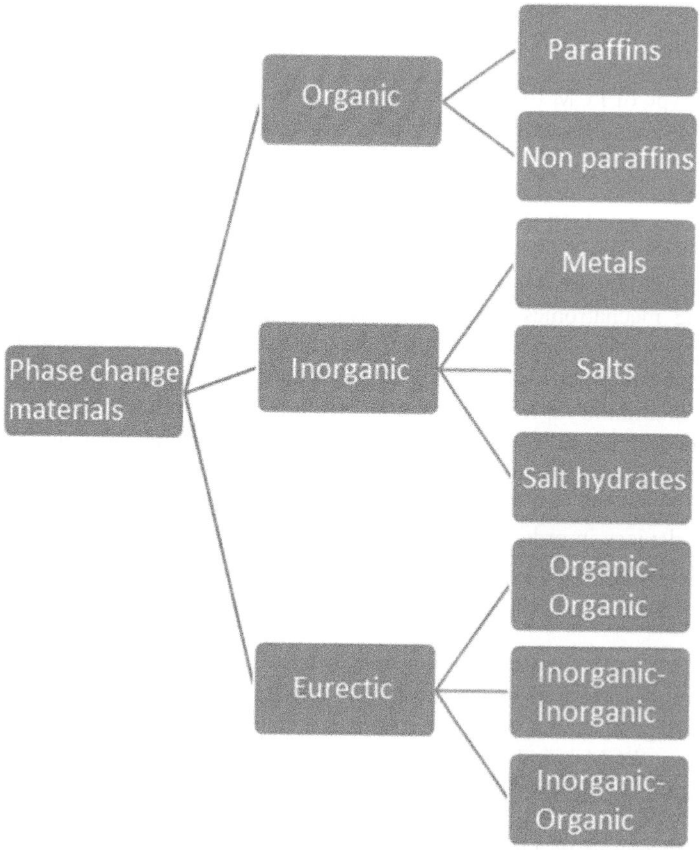

Figure 3.2. *PCM classification. For a color version of this figure, see www.iste.co.uk/ledoux/heat4.zip*

The current challenges for these innovative materials are to strengthen and improve certain properties, such as:

– more precise definition of the melting/solidification range;

– improvement of PCM reaction to fire;

– characterization of potentially emitted products (toxicity);

– cycle durability improvement (characterization and improvement of the number of melting/solidification cycles that the materials can be subjected to without degradation);

– etc.

In addition to these improvement projects, other research projects were launched in order to study the coupling of the PCM with super insulating materials (VIP).

All the previously described characteristics brought the research to the point of using this type of PCM for electronic boards in the field of aerospace, knowing that these boards are subjected to high temperatures during their operation.

3.2. PCM for cooling mechatronics devices

This section presents a system for efficient thermal dissipation based on a PCM for cooling mechatronics devices. In order to maintain the temperature of such devices below the critical temperature, the salt hydrate and wax are used as PCMs absorbing the thermal heat released by electronic components. The PCM is filled inside the aluminum fin HS. Natural convection is used in the present study as a heat transfer mode for improving the operating time of the HS and reaching the critical temperature setting. The heat is studied under three different heat fluxes of 1,250, 2,500 and 5,000 [W/m^2]. The PCM-based HS can store the heat generated by the electronic devices in order to delay its temperature peak. This section compares the distribution of temperature of an HS without and with PCM for various levels of the heat flux.

In most cases, standard cooling methods are not sufficient. In this context, PCM-based HS are part of the most well-known cooling systems. This system is very advantageous and necessary in order to increase the performances of electronic components. PCM-based HS are used in various fields: laptops, aerospace, smartphones, etc. [DEB 21].

The selection of PCMs for a specific application relies on the maximal operating temperature of the device, meaning that the melting temperature of PCMs must be below the maximal operating temperature of the component. In general, the global maximal temperature authorized for such a component should not exceed 85°C to 120°C.

Moreover, PCMs are characterized by their high melting heat, which makes it possible to reach a high storage density. The heat dissipated by the component can be absorbed by PCMs and then dissipated, taking advantage of the thermo-physical properties of PCMs. Then, PCM-based HS can efficiently store the heat dissipated by the components in order to delay its maximal temperature.

Generally, copper and aluminum are the most commonly used metals for improving the global thermal response of the radiator. Even though the thermal conductivity of aluminum is below that of copper, its use is preferable, given the low density, the corrosion resistance and the easy machinability.

The literature includes many studies elaborated for the study of the thermal evolution of various PCM-based radiator configurations. Many chemical substances can be used as PCMs. Various means to improve thermal conductivity were tested, such as the addition of additives that have good thermal conductivity, such as aluminum powder, nanomaterials, carbon fiber or composite chips, integrating PCMs in porous metals or in matrices of expanded graphite or in additional fins or metallic spheres.

3.2.1. *Description of heat transfer and PCM behavior*

The exchange of thermal energy between physical systems is known as heat transfer. It presents the heat transfer from the highest to the lowest temperatures. Three heat transfer modes can be defined: conduction, convection and radiation.

3.2.1.1. *Conduction*

Thermal conductivity is a direct transfer of energy in a material. This thermal transfer mode is specific to solids (wood, metal, etc.).

The heat flux circulates always from the hottest area to the coldest area. In this study, conduction can be represented by the heat transfer between the electronic device, known as the heat source, and the cooling system, known as the receiver. Mathematically, thermal conduction can be presented by the Fourier law, as follows:

$$Q = -\lambda . \nabla T \qquad [3.1]$$

$$\vec{\nabla} T = (\frac{\partial T}{\partial x}\vec{\imath} + \frac{\partial T}{\partial y}\vec{\jmath} + \frac{\partial T}{\partial z}\vec{k}) \quad [3.2]$$

where Q is the heat transfer rate, λ is the thermal conductivity, and T is the temperature.

3.2.1.2. *Convection*

The convection phenomenon relates to the heat transfer that occurs in circulating liquid or gaseous fluids.

There are two types of convection: forced and natural. In forced convection, the fluid is set into motion by the action of external forces, such as those exerted by a fan or pump.

Finally, thermal convection can be defined as the transfer of energy between the surface of an object and a fluid in motion (liquid, gas, air, etc.). In mathematical terms, thermal convection is defined by the following equation:

$$Q = -h.(T_s - T_\infty) \quad [3.3]$$

where h is the convective heat transfer coefficient, and T_∞ and T_s are respectively the fluid temperature and the surface temperature.

3.2.1.3. *Radiation*

Thermal (or diffusive) radiation is an emission of electromagnetic radiation by a hot object. The electromagnetic energy received by the system is absorbed and converted into thermal energy (heat). In fact, there is cooling by radiation when a hot surface is surrounded by cold surfaces that depend on the emissivity of the material denoted by "$\mathcal{E}$". The emissivity $\mathcal{E}$ is a dimensionless number. It reflects the capacity of a material to emit energy by radiation. The more the material radiates heat, the closer to 1 this emissivity $\mathcal{E}$ is.

It can be defined by:

$$Q = \sigma\varepsilon(T_{sur}^4 - T_\infty^4) \quad [3.4]$$

where σ is the Stephan–Boltzmann constant, ε = 0.97 is the external emissivity, T_{sur} is the surface temperature of the HS sides, and T_∞ is the temperature at infinity.

3.2.2. PCM behavior

A PCM is a substance with a high melting heat and which, by turning into a liquid or into a solid at a certain temperature, can store or release large amounts of energy. Heat is absorbed when the material changes its state from solid into liquid, and it is released when it passes from the liquid state to the solid state. The high latent heat of PCMs is efficient for absorbing heat and slowing down the increase in temperature of electronic components. Its integration into a cooling system will therefore be ideal for periodically operating devices. When the latent heat of the PCM is exhausted, heat is still generated and the temperature increases up to a steady regime. The heat released by the electronic component is led by the PCM to the outer edge of the HS and is evacuated by the natural convection of air. Since the efficiency of this natural convection is not high, additional dissipation techniques must be deployed for the operation of the electronic equipment.

The energy absorption by the PCM triggers its transformation from solid into liquid, and its dissipation triggers its transformation from liquid into solid. The thermo-physical properties of PCMs are assumed independent of temperature. The fraction of material in liquid state is estimated at 0 for a complete solid, 1 for a complete liquid and between 0 and 1 for a pure PCM. The three-dimensional energy equations can be written as follows:

$$\frac{\partial}{\partial t}(\rho C_P T) = \frac{\partial}{\partial x}\left(\lambda\frac{\partial T}{\partial x}\right) + \frac{\partial}{\partial y}\left(\lambda\frac{\partial T}{\partial y}\right) + \frac{\partial}{\partial z}\left(\lambda\frac{\partial T}{\partial z}\right) + S_h \qquad [3.5]$$

where ρ is the density, C_P is the specific heat at constant pressure, and S_h is the source term.

The enthalpy of the material is the sum of the enthalpy h and sensible and latent heat ΔH, such that:

$$H = h + \Delta H \qquad [3.6]$$

with:

$- h = h_{ref} + \int_{T_{ref}}^{T} C_p dT$

$- \Delta H = \beta . L$

$$- \beta = \begin{Bmatrix} 0 & if & T < T_{sol} \\ \frac{T - T_{sol}}{T_{liq} - T_{sol}} & if & T_{sol} < T < T_{liq} \\ 1 & if & T > T_{liq} \end{Bmatrix}$$

where h_{ref} is the reference enthalpy, T_{ref} is the reference temperature, T_{sol} is the solid temperature of PCM, T_{liq} is the liquid temperature of PCM, and β is the liquid of PCM that corresponds to the quantity of liquid at the same time with respect to the total quantity of PCM. The enthalpy of the material can therefore be written as follows:

$$H = h_{ref} + \int_{T_{ref}}^{T} C_p dT + \beta . L \quad [3.7]$$

3.3. Numerical model

3.3.1. *Description of the 3D model*

A numerical study was conducted on the HS, without and with PCM and by using two different types of PCMs. Figure 3.3 presents the 3D model of a heat sink.

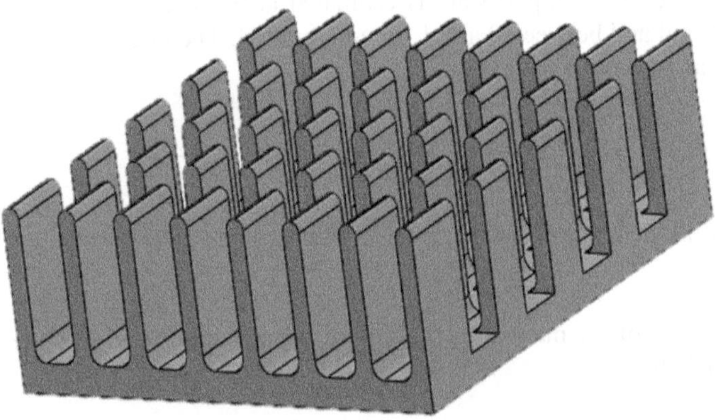

Figure 3.3. *3D model of a heat sink*

This HS is made from aluminum and its grid is presented in Figure 3.4.

Figure 3.5 presents the various modes of heat transfer between the HS and the external environment. The bottom of the HS is subjected to the thermal conduction caused by the heating of the electronic component. The HS was studied under three different heat fluxes of 1,250, 2,500 and 5,000 [W/m^2] at an ambient temperature of 297 K.

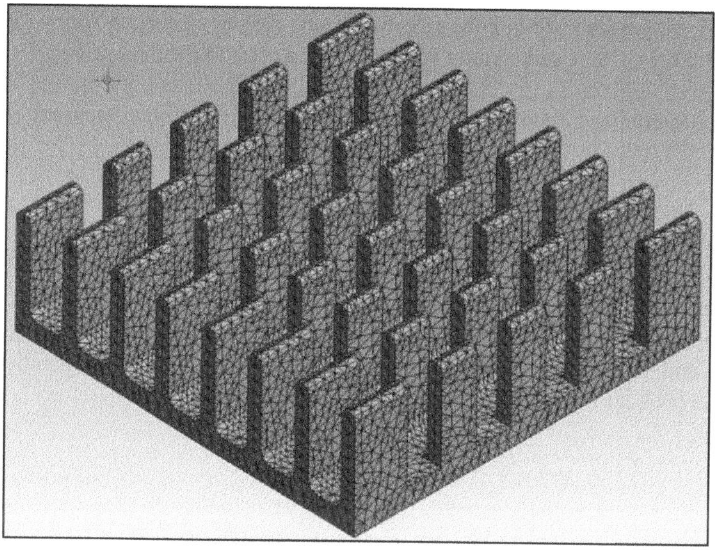

Figure 3.4. *Grid of an HS without PCM*

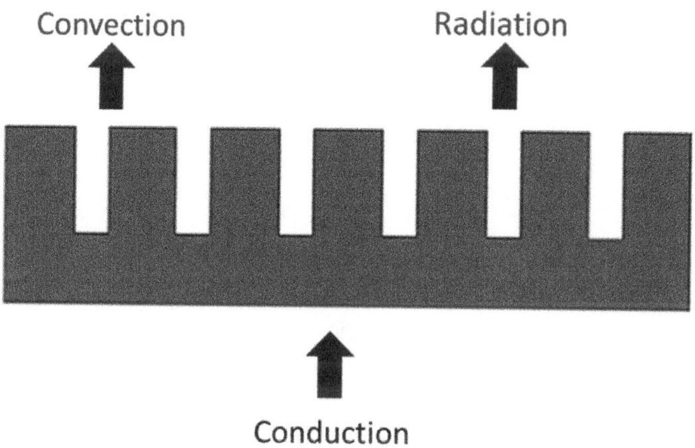

Figure 3.5. *Modes of thermal transfer between the HS and the external environment*

The heat generated by the electronic component is transferred by conduction through the HS. It is then transmitted by convection (natural convection is considered in this study) and by radiation to the ambient air.

Considering the surface of the HS, these heat fluxes correspond to 1,250, 2,500 and 5,000 [W/m^2]. The same study was conducted with PCM integration.

The problem was then modeled and solved using a finite element software, ANSYS.

3.3.2. *Heat sink without PCMs*

This section studies the evolution of temperature for an HS subjected to natural convection and radiation. The model is divided into 48,908 elements for a size equal to 2 mm and 91,022 nodes. As previously mentioned, the HS was studied under three different heat fluxes: 1,250, 2,500 and 5,000 [W/m^2].

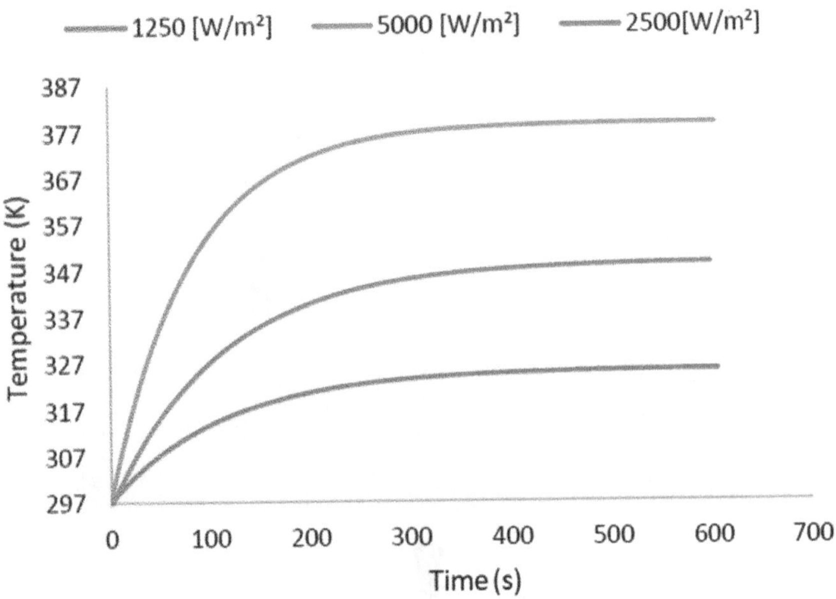

Figure 3.6. *Transient variation of the HS temperature under the three heat fluxes. For a color version of this figure, see www.iste.co.uk/ledoux/heat4.zip*

Figure 3.6 shows the various temperature profiles of the HS under these three fluxes selected for this study. As expected, the temperature at the equilibrium state increases with the increase in the heat flux, as is clearly shown in Figure 3.6. The temperature at the state of equilibrium then becomes respectively 347 K for 2,500 [W/m^2] and 380 K for 5,000 [W/m^2] instead of 325 K for 1,250 [W/m^2].

At the end of the simulation and after reaching the state of equilibrium, the temperature profile of the HS is below 1,250 [W/m^2]. We present this result in Figure 3.7.

The temperature reaches its maximal value at the center of the HS. It progressively decreases with the distance from the center and up to the top of the fins where convection has a significant influence.

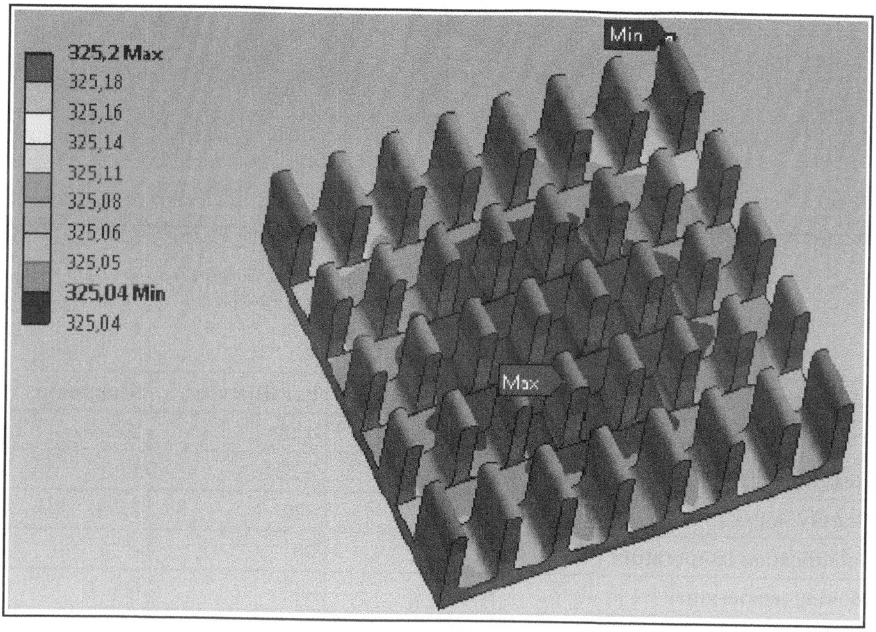

Figure 3.7. *Distribution of temperature in the HS. For a color version of this figure, see www.iste.co.uk/ledoux/heat4.zip*

3.3.3. *Heat sink with PCMs*

Let us now consider that the cavities of the HS are filled by the PCM, as presented in Figure 3.8.

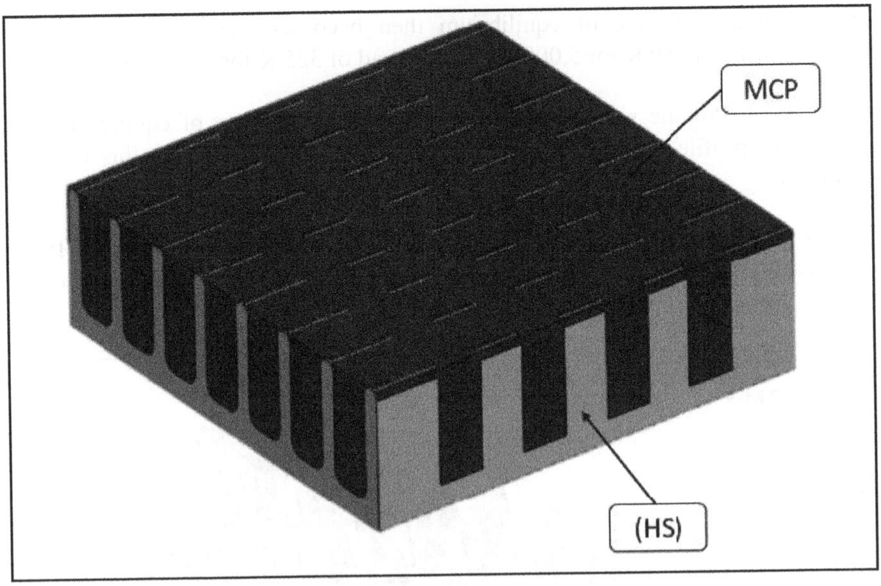

Figure 3.8. *HS model with PCM. For a color version of this figure, see www.iste.co.uk/ledoux/heat4.zip*

Materials	Salt hydrate	Paraffin wax	Aluminum
Thermal conductivity [W/m. K]	0.6	0.15	202.4
Heat capacity [J/kg. K]	2,000	2,200	870
Density [kg/m^3]	1,500	820	2,700
Solidification temperature [°C]	27	30	–
Melting temperature [°C]	32	40	606.4
Latent heat [kJ/kg]	200	190	–

Table 3.1. *Thermo-physical properties of PCMs*

This collection is modeled by a finite element software. Throughout the phase change of the PCM, the volume is considered constant and the molten PCM does not get out of the HS. Two types of PCMs are used here: paraffin wax and salt hydrate.

The material properties of each component of the set are mentioned in Table 3.1.

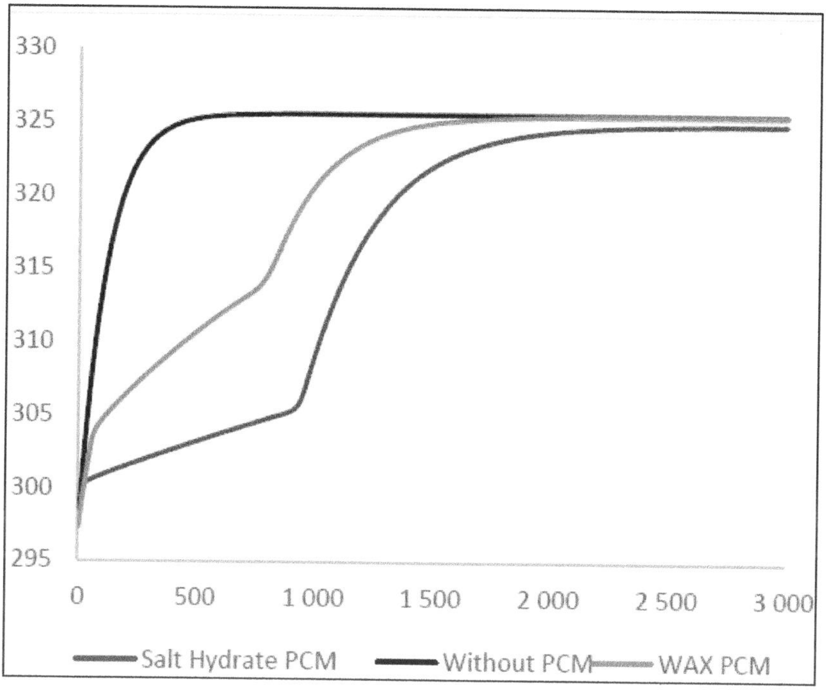

Figure 3.9. *Transient variation of the HS temperature under the three heat fluxes for heat sinks with and without PCM. For a color version of this figure, see www.iste.co.uk/ledoux/heat4.zip*

In order to study the effect of PCMs, Figure 3.9 presents the variation of temperature with and without PCM.

Figure 3.9 shows that, at first, the temperature increases similarly for heat sinks with and without PCM until reaching the melting temperature of the PCM. Then, the salt hydrate preserves its temperature while continuing to increase, in the HS without PCM, until reaching the temperature at the equilibrium state (325 K). The temperature curves with PCM indicating three distinct regions: the solid phase, the latent heating phase and a liquid phase.

Figure 3.10 shows the transient variation of the temperature of the HS with PCM under three heat fluxes. It is important to note that the PCM used in this section is salt hydrate.

We note that, similar to the case without PCM, the temperature at equilibrium state increases with the increase in the heat flux. Moreover, the time required for the latent heating phase increases when the thermal power decreases. For example, for the 1,250 [W/m^2] flux, the time required for full melting of PCM is around 1,000 s, while it is about 500 s and 200 s for the fluxes of 2,500 [W/m^2] and 5,000 [W/m^2] respectively.

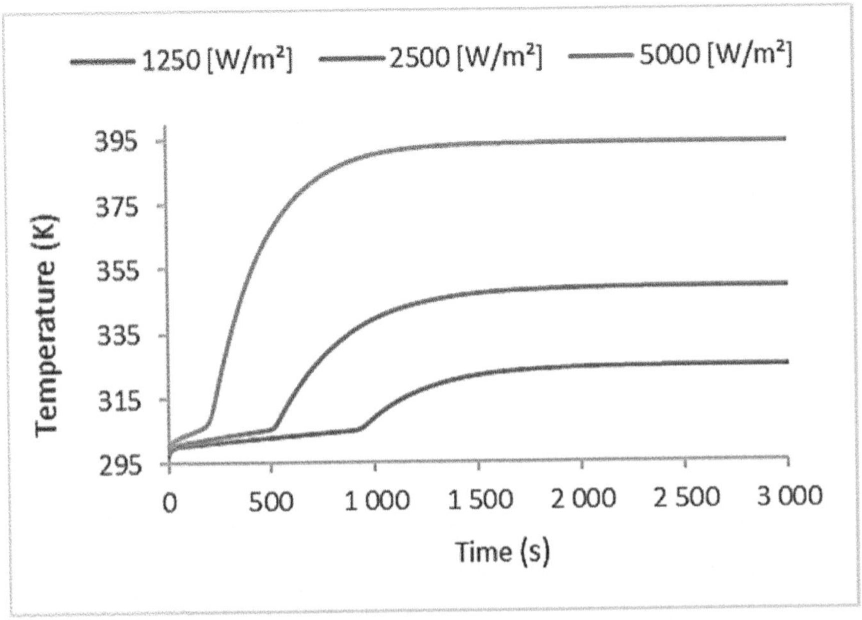

Figure 3.10. *Transient variation of the temperature of heat sinks with PCM under three fluxes of heat. For a color version of this figure, see www.iste.co.uk/ledoux/heat4.zip*

Let us now assume that the HS is subjected to a loading composed of two phases:

– a heating phase 0 → 1,500 s (Q = 1,250 W/m^2);

– a cooling phase 1,500 s → 3,000 s (Q = 0).

Figure 3.11 presents the results found for the evolution of temperature for a heating and cooling of the HS without PCM (black curve) and with PCM (red curve for salt hydrate and blue curve for the wax). It can be noted that, for the whole loading cycle (heating + cooling), the maximal temperature reached by the HS with salt hydrate is well below the maximal temperature measured for the HS without PCM and with wax. Indeed, at 1,500 s, the stabilization temperature is not yet reached by the salt hydrate. Or, for the HS without PCM, the stabilization temperature is reached after t = 600 s.

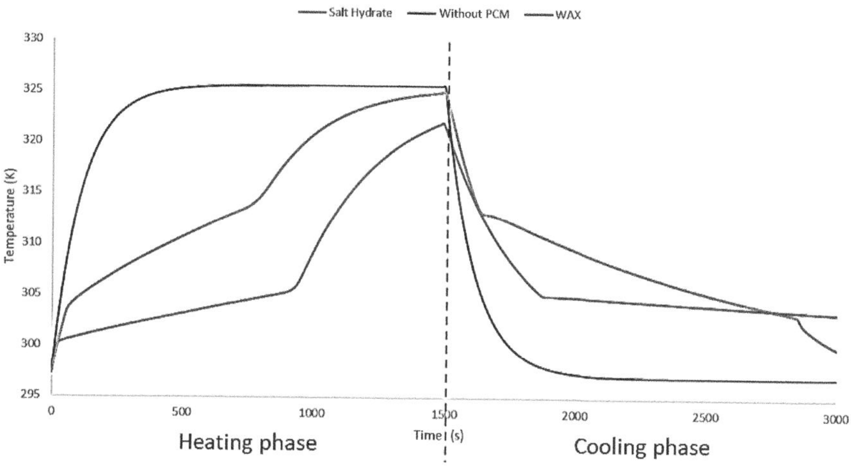

Figure 3.11. *Evolution of temperature for the heating and cooling cycle of the HS without and with PCM. For a color version of this figure, see www.iste.co.uk/ledoux/heat4.zip*

3.4. Conclusion

A PCM-based HS designed for various levels of power was numerically studied. The results show that the inclusion of PCM in the cavities of radiators increases the cooling performances with respect to the cases without PCM when the input power level is relatively high (in this study, for Q > 1,250 [W/m^2]).

After the simulations of performances, it can be concluded that: the static temperature of the system to be studied increases when the heat flux increases. PCMs also allow the delay of the maximal temperature by absorbing heat, which leads to better long-term performances due to the intermittent use of the component.

4

Electro-Thermo-Mechanical Modeling of Systems

4.1. Introduction

The main focus of this chapter is the finite element modeling of electronic boards subjected to electrical, thermal and vibrational stress.

It presents the coupling between the three domains of a mechatronics structure, namely electrical, thermal and mechanical. Two types of coupling are presented. The first is a strong coupling; it uses finite elements having all the degrees of freedom required for an electro-thermo-mechanical study. The second one is a weak coupling; it uses sequential computation to decouple the three physical phenomena.

This method is used in two applications. The first one is an electronic board of an engine control unit. The second one is a radar power amplifier. Knowledge on the mechanical behavior of electronic boards requires modeling several physical phenomena. The multiphysics modeling presented takes into consideration the interdependences and the interactions between various physical phenomena of electrical, thermal and vibrational nature.

Electronic boards are generally subjected to severe and varied stresses. These boards are often made up of printed circuits and electronic components (capacitors, resistors, transistors, etc.) brazed on these circuits. Indeed, the electric pulses due to the internal operation of electronic boards generate thermal loadings inducing mechanical deformations that may exceed the desired thresholds. Moreover,

For a color version of all the figures in this chapter, see www.iste.co.uk/ledoux/heat4.zip.

embedded mechatronic systems are subjected to vibrations that may impact the state of stresses in the solder joints leading to fatigue damage. Hence, knowing the mechanical response of electronic boards is an essential stake for the electronic industry, since these boards fulfill strategic functions as in the case of flight controls. Multiphysics modeling aims to consider the interdependences and the interactions between various physical phenomena, such as electrical, thermal, mechanical and vibrational.

4.2. Theory of electro-thermo-mechanical coupling

As a first step, the equations related to the thermal and electrical problem are determined, and then the types of boundary conditions are examined, in order to solve the mathematical equations describing the physical phenomena of electro-thermo-mechanical coupling.

4.2.1. *Electro-thermal phenomena*

Electro-thermal phenomena correspond to heat generation due to a current flowing through a body. This is known as the Joule effect. A parallel can be drawn between the thermal and electrical phenomena, as the laws operate similarly. There are several ways to electrically heat a part: by conduction, induction or by electric arcs. This study focuses on the effects of electrical conduction, as the purpose of our model is to observe the behavior of our components when they are under voltage. Moreover, since the electro-thermo-mechanical coupling already requires a lot of computation time, the heat effect on the voltage will not be added. The equations governing the Joule effect by electric loading are as follows:

– ***Ohm's law***

According to which

$$U = RI \qquad [4.1]$$

where U is the voltage in Volts, R is the resistance in Ohms, and I is the current intensity in Ampères.

$$P = UI = RI^2 \qquad [4.2]$$

where P is the power expressed in Watts for a purely resistive system loaded with direct current.

– Electrical conduction

There are two types of electrical conduction: direct or indirect. In both cases, the energy needed to heat the part is generated by the Joule effect.

– Direct conduction: the current flows through the part to be heated.

– Indirect conduction: the current flows through a heating resistor. This brings heat to the part to be heated.

In our case, voltage is directly applied to the part. This is the case of direct conduction, which is governed by the equations as follows:

– Neglecting the losses:

$$P = mC_p \frac{dT}{dt} \quad [4.3]$$

hence

$$T = T_0 + \frac{P}{mC_p}\mathrm{t} \quad [4.4]$$

where T_0 is the initial temperature of the part.

– If the losses are not neglected:

$$P - \Phi = P - \mathrm{h_f}S(T - T_0) = mC_p \frac{dT}{dt} \quad [4.5]$$

With Φ the heat flux in Watts:

$$\Phi = \mathrm{h_f}S(T - T_0) \quad [4.6]$$

hence

$$\left(\frac{mC_p}{\mathrm{h_f}S}\right)\frac{dT}{dt} + T = \frac{P}{\mathrm{h_f}S} + T_0 \quad [4.7]$$

where:

– h_f is the exchange coefficient in $W.m^{-2}.K^{-1}$;

– $\frac{dT}{dx}$ is the temperature gradient in Km^{-1};

– Φ is the heat flux in Watts;

– dt is the elementary time in s;

– dt is the elementary time in s.

4.2.2. *Numerical formulation of the electro-thermo-mechanical coupling*

A mathematical formulation was recently proposed for the problem with coupled electrical and thermal boundary conditions. Based on this formulation of the problem with continuous boundary conditions, using a finite element approach, a numerical formulation of the electro-thermo-mechanical coupling can be derived.

4.2.2.1. *Electro-thermal coupling*

The constitutive equations for the electro-thermal coupling are [MAK 16]:

$$\{q\} = [\Pi]\{J\} - [K]\{\nabla T\} \quad [4.8]$$

$$\{J\} = [\sigma]\{E\} - [\alpha]\{\nabla T\} \quad [4.9]$$

Replacing $[\Pi]$ with T $[\alpha]$, the following two equations are obtained:

$$\{q\} = [T(\alpha)]\{J\} - [K]\{\nabla T\} \quad [4.10]$$

$$\{J\} = [\sigma]\{E\} - [\alpha]\{\nabla T\} \quad [4.11]$$

where:

– $[\Pi]$ is the matrix of Peltier coefficients = $T[\alpha]$;

– T is the absolute temperature;

– $\{q\}$ is the heat flux density;

– $\{J\}$ is the density of electric current;

– $\{\nabla T\}$ is the temperature gradient;

– $\{E\}$ is the electric field;

– $[\alpha] = \begin{bmatrix} \alpha_{xx} & 0 & 0 \\ 0 & \alpha_{yy} & 0 \\ 0 & 0 & \alpha_{zz} \end{bmatrix}$ is the matrix of Seebeck coefficients;

– $[K] = \begin{bmatrix} K_{xx} & 0 & 0 \\ 0 & K_{yy} & 0 \\ 0 & 0 & K_{zz} \end{bmatrix}$ is the matrix of thermal conductivities evaluated at $\{J\} = \{0\}$;

– $[\sigma] = \begin{bmatrix} 1/\rho_{xx} & 0 & 0 \\ 0 & 1/\rho_{yy} & 0 \\ 0 & 0 & 1/\rho_{zz} \end{bmatrix}$ is the matrix of electrical conductivities evaluated at $\{\nabla T\} = \{0\}$;

with:

K_{xx}, K_{yy}, K_{zz} being the thermal conductivity;

ρ_{xx}, ρ_{yy}, ρ_{zz} being the coefficients of resistivity;

α_{xx}, α_{yy}, α_{zz} being the Seebeck coefficients.

By applying the principle of virtual works to the equations of heat flux, the following matrix form is obtained [MAK 15]:

$$\begin{bmatrix} [C^t] & [0] \\ [0] & [C^v] \end{bmatrix} \begin{Bmatrix} \{\dot{T}\} \\ \{\dot{V}\} \end{Bmatrix} + \begin{bmatrix} [K^t] & [0] \\ [K^{vt}] & [K^v] \end{bmatrix} \begin{Bmatrix} \{T\} \\ \{V\} \end{Bmatrix} = \begin{Bmatrix} \{Q\} + \{Q^p\} \\ \{I\} \end{Bmatrix} \quad [4.12]$$

4.2.2.2. *Thermo-mechanical coupling*

Thanks to the equations of thermo-elasticity [MAK 15], the deformations and the density of entropy can be defined by:

$$\{\varepsilon\} = [D]^{-1}\{\sigma\} + \{\alpha\}\Delta T \quad [4.13]$$

$$\{S\} = \{\alpha\}^T \{\sigma\} + \frac{\rho C_P}{T_0}\Delta T;\ \Delta T = T - T_{ref} \quad [4.14]$$

where:

– $\{\varepsilon\}$ is the total deformation vector = $[\varepsilon_x\ \varepsilon_y\ \varepsilon_z\ \varepsilon_{xy}\ \varepsilon_{yz}\ \varepsilon_{xz}]^T$;

– S is the *density of entropy*;

– $\{\sigma\}$ is the stress vector = $[\sigma_x\ \sigma_y\ \sigma_z\ \sigma_{xy}\ \sigma_{yz}\ \sigma_{xz}]^T$;

– T is the current temperature;

– T_0 is the absolute reference temperature = $T_{ref} + T_{off}$;

– T_{ref} is the reference temperature;

– T_{off} is the offset temperature;

– $\{\alpha\}$ is the vector of thermal expansion coefficients = $[\alpha_x\ \alpha_y\ \alpha_z\ 0\ 0\ 0]^T$;

– ρ is the density;

– C_p is the specific heat under constant pressure;

– [D] is the elastic rigidity matrix, for which the inverse matrix is defined by the following equation:

$$[D]^{-1} = \begin{bmatrix} 1/E_x & -\vartheta_{xy}/E_x & -\vartheta_{xz}/E_x & & & \\ -\vartheta_{yx}/E_y & 1/E_y & -\vartheta_{yz}/E_y & & 0 & \\ -\vartheta_{zx}/E_z & -\vartheta_{zy}/E_z & 1/E_z & & & \\ & & & 1/G_{xy} & 0 & 0 \\ & 0 & & 0 & 1/G_{yz} & 0 \\ & & & 0 & 0 & 1/G_{xz} \end{bmatrix}$$

where:

$E_{x,y,z}$ is Young's modulus;

$v_{xy,xz,yz}$ is Poisson's ratio;

$G_{xy,xz,yz}$ is the shear modulus.

According to the second law of thermodynamics, for a reversible process, the *density of entropy* S (equation [4.24]) is replaced by the density of heat Q:

$$Q = T_0\,S$$

This yields the following:

$$\{\sigma\} = [D]\{\varepsilon\} - \{\beta\}\Delta T \qquad [4.15]$$

$$Q = T_0\{\beta\}^T\{\varepsilon\} + \rho C_V \Delta T \qquad [4.16]$$

where:

– $\{\beta\}$ is the vector of thermo-elastic coefficients = [D] {α};

– C_V is the specific heat at constant volume = $C_P - \frac{T_0}{T}\{\alpha\}^T\{\beta\}$.

Substituting Q (equation [4.27]) in the heat flux equation (equation [4.9]):

$$\frac{\delta Q}{\delta t} = T_0\{\beta\}^T\frac{\delta\{\varepsilon\}}{\delta t} + \rho C_V\frac{\delta\{\Delta T\}}{\delta t} - [K]\nabla^2\mathrm{T} \qquad [4.17]$$

By applying the principle of virtual works of stresses, and the equation of heat flux conservation coupled by the thermo-elasticity equations, the following matrix form is obtained:

$$\begin{bmatrix} [M] & [0] \\ [0] & [0] \end{bmatrix} \begin{Bmatrix} \{\ddot{u}\} \\ \{\ddot{T}\} \end{Bmatrix} + \begin{bmatrix} [C] & [0] \\ [C^{tu}] & [C^t] \end{bmatrix} \begin{Bmatrix} \{\dot{u}\} \\ \{\dot{T}\} \end{Bmatrix} + \begin{bmatrix} [K] & [K^{ut}] \\ [0] & [K^t] \end{bmatrix} \begin{Bmatrix} \{u\} \\ \{T\} \end{Bmatrix} = \begin{Bmatrix} \{F\} \\ \{Q\} \end{Bmatrix} \qquad [4.18]$$

4.2.2.3. *Electro-thermo-mechanical coupling*

Taking [4.22] and [4.29] into account, this finally leads to writing the matrix equation of the electro-thermo-mechanical coupling in the form:

$$\begin{bmatrix} [M] & [0] & [0] \\ [0] & [0] & [0] \\ [0] & [0] & [0] \end{bmatrix} \begin{Bmatrix} \{\ddot{u}\} \\ \{\ddot{T}\} \\ \{\ddot{V}\} \end{Bmatrix} + \begin{bmatrix} [C] & [0] & [0] \\ [C^{tu}] & [C^t] & [0] \\ [0] & [0] & [C^v] \end{bmatrix} \begin{Bmatrix} \{\dot{u}\} \\ \{\dot{T}\} \\ \{\dot{V}\} \end{Bmatrix} + \begin{bmatrix} [K] & [K^{ut}] & [0] \\ [0] & [K^t] & [0] \\ [0] & [K^{vt}] & [K^v] \end{bmatrix} \begin{Bmatrix} u \\ \{T\} \\ \{V\} \end{Bmatrix} = \begin{Bmatrix} \{F\} \\ \{Q\} \\ \{I\} \end{Bmatrix} \qquad [4.19]$$

where:

$[K^t] = [K^{tb}] + [K^{tc}]$;

$\{F\} = \{F^{nd}\} + \{F^{pr}\} + \{F^{ac}\}$;

$\{Q\} = \{Q^{nd}\} + \{Q^g\} + \{Q^c\} + \{Q^j\} + \{Q^p\}$.

4.3. Finite element simulation of the electro-thermo-mechanical behavior

Electro-thermo-mechanical physical phenomena are mathematically described by partial differential equations (PDEs). The lack of an analytical solution to these PDEs led to the development of methods of approximation such as the finite element method (FEM). The FEM involves the division of the continuous domain into a certain number of elements with simple geometries, known as cells. The set of cells is known as mesh. The discretization process makes it possible to replace the continuous medium with a discrete medium. According to the FEM, the solution of the PDE is approximated using the discretization of its variational formulation. This discretization can be used to replace the variational formulation of the PDE with a matrix formulation. The solution of the PDE is approximated by determining the nodal solutions of the discretized domain. Hence, the sought-for field is approximated and reconstructed by an interpolation at nodes by considering all the cells of the domain.

Moreover, the advantage of using the FEM resides in the possibility of coupling various physical phenomena and behaviors of materials. In other terms, electrical, thermal and mechanical behaviors of the studied structure can be obtained. In fact, there are two types of coupling of physical phenomena: strong coupling and weak coupling. The first one uses a finite element, having all the degrees of freedom required for an electro-thermo-mechanical study. The second one involves decoupling the three physical phenomena, using a specific finite element for each type of physics. The coupling between various physics is iterative. Figure 4.1 presents the diagram for electro-thermo-mechanical modeling.

4.3.1. *Strong coupling of the electro-thermo-mechanical modeling*

The strong coupling of the electro-thermo-mechanical modeling involves the simultaneous resolution of various PDEs in order to take into account the interactions between the various physical phenomena, namely the heat gradient induced by the passage of direct currents and the mechanical deformation due to the heat gradient. The simultaneous resolution of the PDEs of these various physics requires the use of a single finite element, having all the properties necessary to transform the PDEs into a single matrix system, to be solved once. The library of ANSYS software is very rich in finite elements. The volume finite element of tetrahedral type, known as SOLID227, can be used to achieve the strong coupling of a multiphysics modeling. This 10-node element can be used in the electro-thermo-mechanical modeling in order to define five degrees of freedom for each node. The three degrees of freedom are required for the mechanical modeling (displacements u_x, u_y, u_z), one degree of freedom for the thermal modeling (temperature T) and one degree of freedom for the electrical modeling (electrical voltage V). However, one of the drawbacks of the strong coupling of the multiphysics modeling is the increase in the size of the matrix systems to be solved, strongly depending on the meshing and on the number of degrees of freedom associated with each node. For example, the electro-thermo-mechanical modeling of a structure discretized by a meshing of 10,000 nodes requires a resolution of a matrix system whose size may reach 50,000 x 50,000.

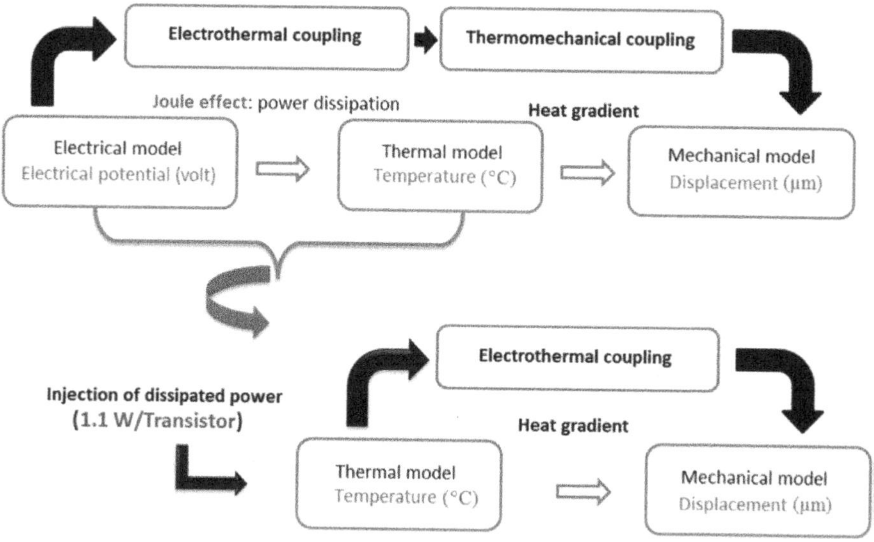

Figure 4.1. *Diagram of the electro-thermo-mechanical modeling*

4.3.2. *Weak coupling of the electro-thermo-mechanical modeling*

The objective of the weak coupling of the electro-thermo-mechanical modeling remains the same as that of the strong coupling, namely to determine the field of electrical voltage, the field of temperature and finally the field of displacements. The weak coupling aims to solve each PDE of a physical phenomenon sequentially. In other terms, the process starts by conducting an electrical analysis taking into consideration the temperature and voltage boundary conditions. The resolution of the electrical problem yields the Joule effect due to the passage of electrical currents. Then, the thermal analysis uses the Joule effect resulted from the electrical analysis as temperature boundary conditions, allowing the calculation of the temperature field. Finally, the mechanical analysis is used to calculate the field of displacements using the previously calculated field of temperature as thermal loading on the studied structure. This process can be repeated until the convergence criteria are met.

4.4. Example of electro-thermo-mechanical simulation of an HBT transistor

This study focuses on an HPA (High Power Amplifier) testing vehicle operating in the ***S*** band (Figure 4.2). The transistors composing the amplifier use GaInP/AsGa heterojunction bipolar technology (HBT). Mastering the design of HPA is essential, as it may generate a strong degradation of the performances of the RF (Radio-Frequency) signal emission module, notably in terms of heat dissipation and of amplitude and phase drift of the RF signal emitted.

The design of the module must be optimal, employing high-performance CAD (Computer-Aided Design) tools. Indeed, the possibility to precisely analyze the electrical, thermal and mechanical behavior before its production limits the malfunction risks. The objective is to obtain an operational module since the first launch. To reach this objective, the process should rely on multiphysics modeling, in which the studied module is subjected to various types (electrical, thermal and mechanical) of stress. Many numerical models have been developed. In other terms, it is a global model enabling the prediction of the electronic board in a global manner, then a local model with a finer meshing allowing the prediction of the life duration or the consideration of complex phenomena of the structural behavior, such as the nonlinear behavior of the materials or the calculation of damage due to the loading cycles.

4.4.1. *Global model of HPA*

This section presents two types of finite element thermo-mechanical simulations. The first part concerns the simulation of the mechanical behavior of the modules while considering the screw tightening effects, the second one is an extension of the first, with the aim of verifying the effect of power injection on HPA behavior.

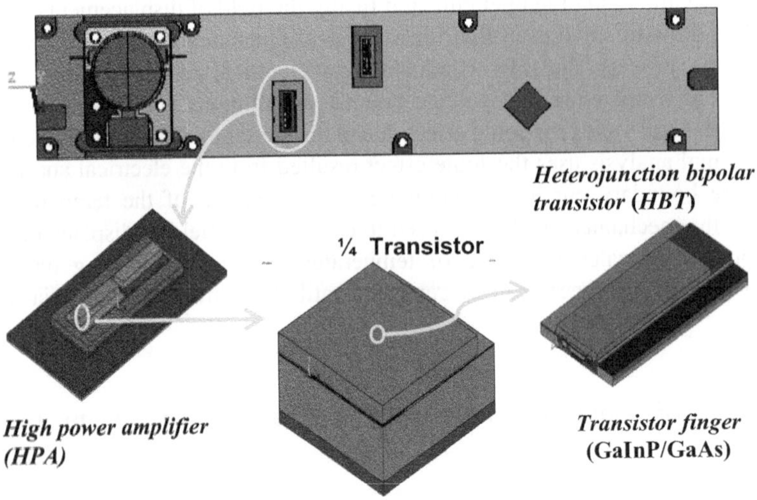

Figure 4.2. *Description of an HPA board*

4.4.1.1. *Numerical modeling of the tightening effects on the behavior of the flange*

The objective is to numerically calculate the out-of-plane mechanical deformations due to the tightening of eight screws of the flanges installed on the Octopack (Figure 4.3). The resulting deformations are compared to the experimental results obtained using optical methods.

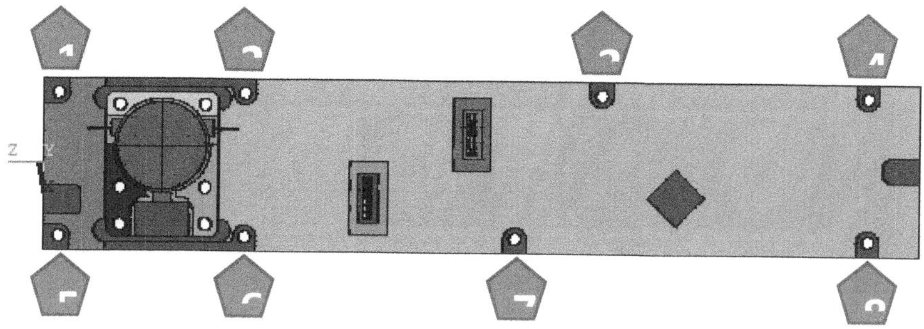

Figure 4.3. *Numbering of eight screws for fixing the flange*

One of the great difficulties of modeling screwed assemblies concerns the introduction of the tightening prestress. Setting up a displacement imposed to the lower nodes of the screw is the best compromise for simulating the tightening prestress. Indeed, during a tightening operation, energizing the screw is obtained by stretching the screw. Consequently, this simple preloading setting is used in all the finite element calculations. To adjust this prestress, an arbitrary displacement is imposed and a calculation without external loading is launched. The average stress observed in the screw can be used to estimate the value of the displacement to be imposed in order to achieve the expected stress.

In order to simplify the modeling of the area of contact between the flange and the Octopack, the latter is not modeled. The effect of the Octopack on the flange is replaced by simple boundary conditions. Considering that all the nodes located in the equivalent zone of contact have no relative displacement (this embedded zone corresponds to the recommendations in the literature at a support diameter that is 3.8 times the diameter of the screw hole, which is 8.7 mm).

Figure 4.4 presents the out-of-plane displacements of the flange during the tightening of eight screws, the maximal values of the displacements are given in each case. The study reveals that four "inner" screws 2, 3, 6 and 7 generate much more significant displacements compared to the screws located at the corners of the board. The results of the numerical model are close to the experimental results obtained by Dan Borza using speckle interferometry techniques (Figure 4.5).

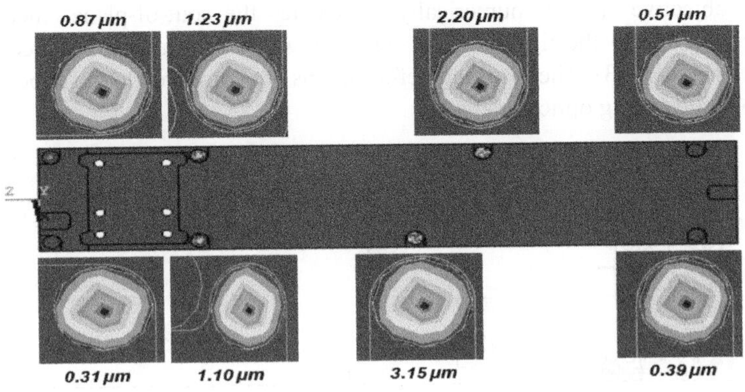

Figure 4.4. *Displacements due to the tightening of eight screws*

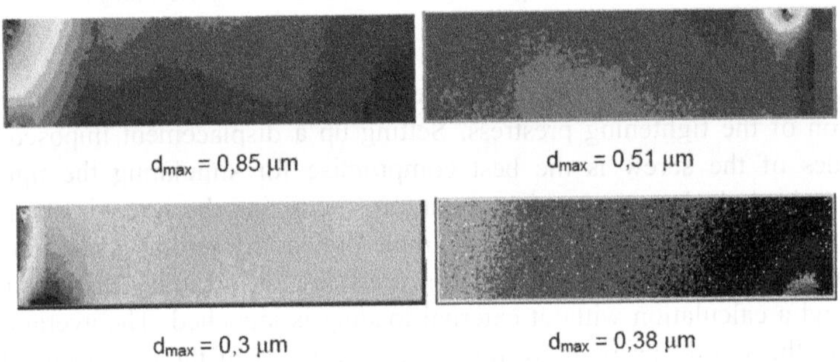

Figure 4.5. *Displacements due to the tightening of outer screws*

4.4.1.2. *Global modeling of the HPA amplifier*

The HPA power amplifier is composed of 12 AsGa/GaInP heterojunction bipolar transistors identical to those previously studied. In order to describe the power amplifier from a system point of view, the transistor model must once more be simplified. Let us recall the transistor materials having a major impact on heat dissipation: AsGa, GaInP and gold (Au). The amplifier model is achieved by integrating the transistors composed of an AsGa power bar, GaInP and gold layers on an equivalent volume of the AsGa substrate (Figure 4.6).

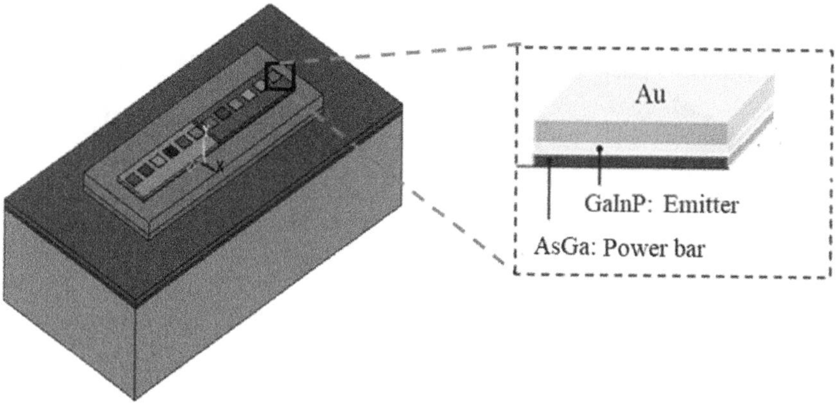

Figure 4.6. *Composition of the simplified finite element model of a power amplifier*

The calculation is conducted in two stages with the ANSYS software. The first stage involves a purely thermal simulation by injecting a heat flux on the active surfaces of the transistors. The ambient temperature is imposed as boundary conditions on the lower surface of the flange. Then, the second stage involves the introduction of a field of temperature as an external loading in the mechanical model for the calculation of mechanical deformations.

In order to evaluate the thermal interaction between the transistors, a set of nonlinear analyses in a steady state was conducted. For example, the influence of transistor T2 on transistor T1 is evaluated considering these active transistors. The "power bar" of T1 and T2 is active; the other transistors are inactive. The thermal interaction of T2 and T1 is evaluated by comparing the value of the maximal temperature in this configuration and the temperature of T1 when it is the only one in active state. Each interaction between transistors is evaluated by this method.

Figure 4.7 shows the distribution of the out-of-plane displacement in the studied power amplifier depending on temperatures, where the different types of thermal couplings are eliminated. The hottest transistor is located at the center of the amplification; its localization at the core of the amplifier does not favor heat evacuation. The thermal interaction between transistors is weak; the maximal interaction on the central transistor is only 0.6°C.

The results have also shown that the component was displaced by 195 nm (195×10^{9}m) out of plane for a temperature difference of 70°C, the results being close to those found experimentally (Figure 4.8).

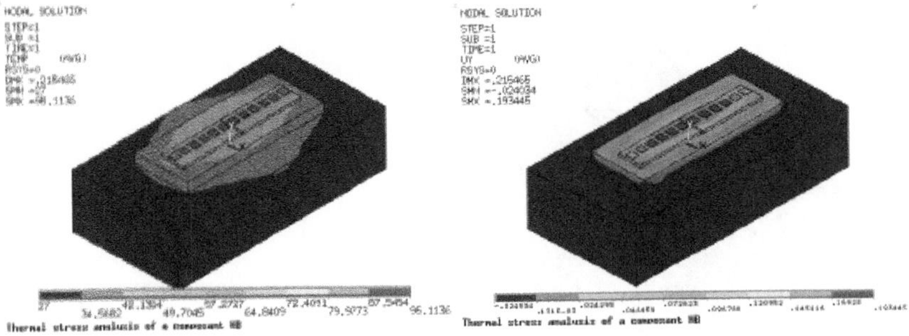

a) Displacement in µm b) Temperature in °C

Figure 4.7. *Distribution of the field of deformation (a) and temperature (b) throughout the studied structure*

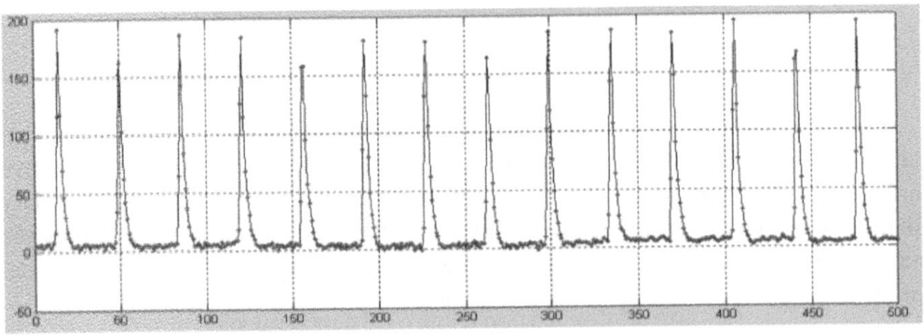

Figure 4.8. *Time history of the out-of-plane displacements in the HPA power amplifier*

4.4.2. *Local model of an HBT transistor*

The considered power amplifier has two amplification stages composed of 12 AsGa/GaInP heterojunction bipolar transistors. A "finite element" model of HBT is achieved in three dimensions. In order to significantly reduce the numerical calculation redundancies, the symmetry properties of the component are taken into account (Figure 4.9) and enable the simplification of the modeling. Furthermore, the contacts between materials are assumed perfect.

The software used imposes a library of predefined elements associated with interpolation functions that allow a coupled analysis of various (electrical and thermal) fields to be studied. The types of elements are adequately assigned according to the shape and dimensions of the volume to be "meshed". The disparity of the considered volume dimensions, notably of fine epitaxial layers at the level of heterojunctions (Figure 4.10) compared to the volume of the AsGa substrate, renders the global resolution of the system difficult with the types of elements proposed by the software.

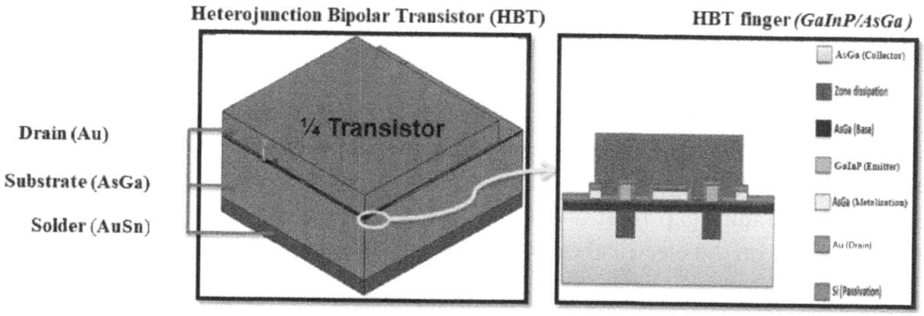

Figure 4.9. *Description of the 3D model of HBT taking into account the symmetry properties*

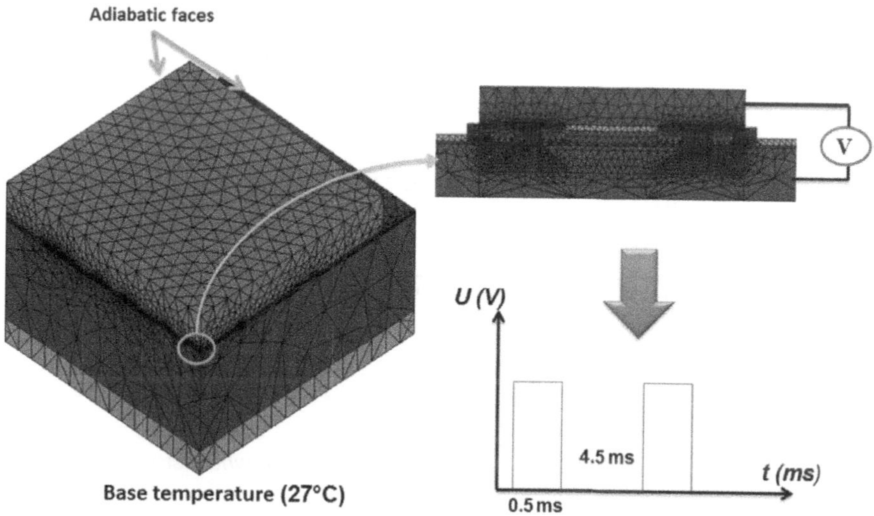

Figure 4.10. *Description of the 3D model of HBT taking into account the symmetry properties*

All these hypotheses are deduced from the knowledge on the origin of heat dissipation. For an HBT, this zone is located in the space load zone (strong electrical field), the place crossed by the density of collector current *Jc* below the fingers of the emitter, in the collector. The current is assumed uniform over all of the fingers, which makes it possible to assign a uniform volume power in the dissipation zone. In the model of our study, this zone is defined by a volume of $120\ \mu m^3$. Once the meshing is achieved (Figure 4.10), the imposed boundary conditions are:

– adiabatic faces;

– base temperature fixed at ambient temperature (27°C).

The distribution of heat over a quarter of the HBT structure is presented in Figure 4.11. The thermal interaction phenomenon appears very clearly: the central fingers have a higher temperature than the most external fingers. Figure 4.11 shows that this temperature gap is of about 15%. This non-negligible value may have an impact, notably on the electrical characteristics of the component (Crunch effect).

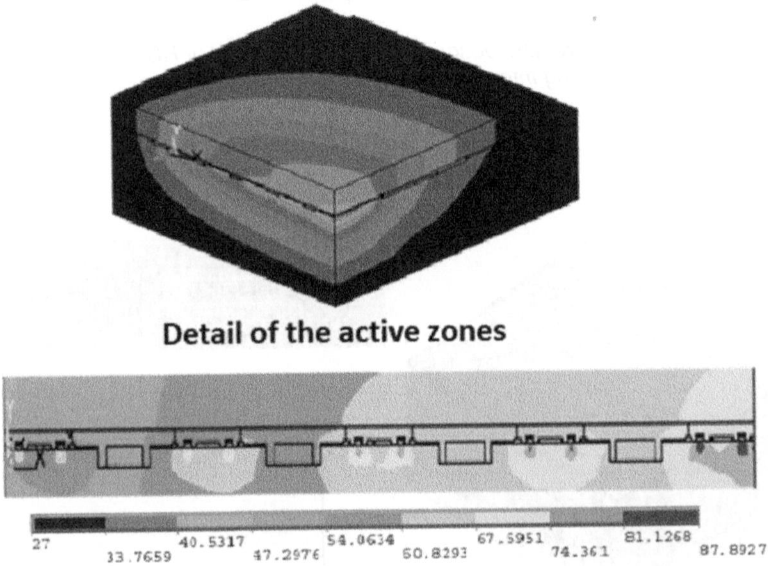

Figure 4.11. *Global distribution of the temperature over the whole studied structure*

Numerical simulations using the FEM in transient regime were conducted by imposing a power step of 0.260 W to the transistor model (Figure 4.10).

Figure 4.12 shows the temperature increase of 10 fingers of emitters of an HBT transistor. It is interesting to note that during the first 3 microseconds, the fingers remain thermally independent, before the occurrence of the interaction phenomenon that generates a marked separation of the temperature curves. Moreover, in one millisecond, it can be noted that the structure does not reach steady state, which is effective after several dozen milliseconds. Figure 4.12 also shows that the farther the finger is considered from the center of the transistor, the less significant the thermal interactions are. Indeed, finger d1 is influenced only on one side, its temperature being lower.

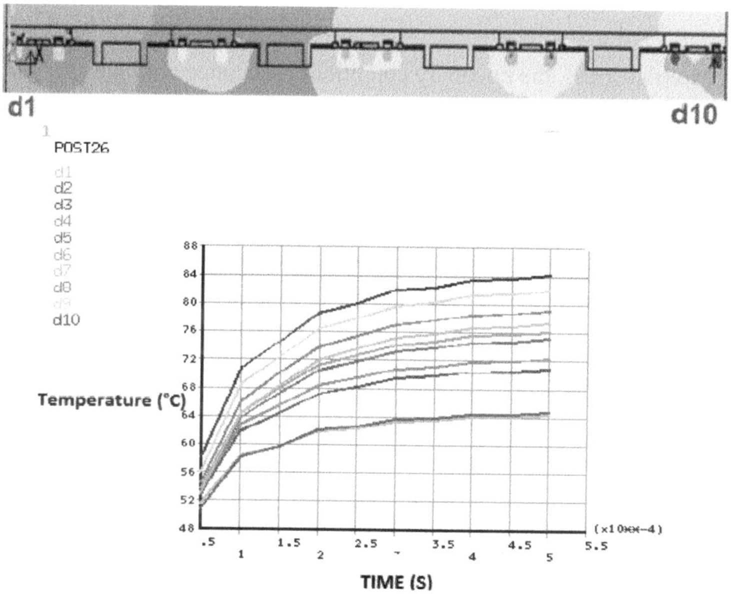

Figure 4.12. *Time evolution of the maximal temperature for each "finger" di of the transistor subjected to a continuous power for a base temperature of 27°C*

4.5. Modal analysis of mechatronic components

As a first step, the equations governing the vibrational behavior of the electronic components are determined. Then, the resulting equations are written under variational form and discretized by the FEM, which makes it possible to obtain matrix systems to be numerically solved in the frequency domain in order to find eigenfrequencies and eigenmodes.

4.5.1. *Writing the equation of the vibration problem*

The considered structures have an elastic, linear, isotropic behavior, lacking stress or initial deformation. In the absence of the volume source, the equation that governs its vibrational behavior is given by:

$$\nabla.\boldsymbol{\sigma} - \rho_s \ddot{\mathbf{u}} = \mathbf{0} \qquad [4.20]$$

ρ_s, $\mathbf{u}$ and $\boldsymbol{\sigma}$ are respectively the density, the field of displacement and the tensor of stresses of the structure.

Denoting by Γ_u the boundaries of imposed displacement and by Γ_f those of imposed external stress, the boundary conditions associated with the structure can be written as:

$$\mathbf{u}\big|_{\Gamma_u} = \bar{\mathbf{u}}, \qquad \boldsymbol{\sigma}.\mathbf{n}\big|_{\Gamma_f} = \bar{\mathbf{f}} \qquad [4.21]$$

For a linear elastic material, stresses and deformations are linked by Hooke's law $(\boldsymbol{\sigma} = \mathbf{D}\boldsymbol{\varepsilon})$. The symmetry of stress and deformation tensors can be used, and stresses and deformations can be represented in the form of a column vector with six components. Hooke's law is then written as follows:

$$\begin{pmatrix} \sigma_{xx} \\ \sigma_{yy} \\ \sigma_{zz} \\ \sigma_{xy} \\ \sigma_{xz} \\ \sigma_{yz} \end{pmatrix} = \frac{E}{(1+\nu)(1-2\nu)} \begin{bmatrix} 1-\nu & \nu & \nu & 0 & 0 & 0 \\ \nu & 1-\nu & \nu & 0 & 0 & 0 \\ \nu & \nu & 1-\nu & 0 & 0 & 0 \\ 0 & 0 & 0 & 1-\nu/2 & 0 & 0 \\ 0 & 0 & 0 & 0 & 1-\nu/2 & 0 \\ 0 & 0 & 0 & 0 & 0 & 1-\nu/2 \end{bmatrix} \begin{pmatrix} \varepsilon_{xx} \\ \varepsilon_{yy} \\ \varepsilon_{zz} \\ 2\varepsilon_{xy} \\ 2\varepsilon_{xz} \\ 2\varepsilon_{yz} \end{pmatrix} \qquad [4.22]$$

Geometry	Modeling	Number of ddls per node	Translation	Rotation
1D	Bar	1	1	–
	Beam	2	1	1
	Axisymmetric shell	3	1	2
2D	Structure in-plane state of stresses	2	2	–
	Structure in-plane state of deformations	2	2	–
	Plate	3	1	2
	Shell	5	3	2

Table 4.1. *Number of ddls corresponding to several models*

Structural mechanics always aims to approximate the geometry, kinematics and the state of stresses in order to simplify the three-dimensional problem, and reduce the number of unknowns (degrees of freedom). This is how simplified models were created, such as the 1D bar or beam elements and the 2D plate or shell elements. Table 4.1 exposes the number of degrees of freedom per node corresponding to several models.

4.5.2. *Variational formulation*

Let us give $\mathbf{u}^*$ a virtual field associated with the structure. Equation [4.21] is integrated over the domain Ω_s. After integration by parts and then application of conditions [4.32], the variational problem involves finding u such that $\left\{\mathbf{u}|_{\Gamma_u} = \bar{\mathbf{u}}\right\}$:

$$\int_{\Omega_s} \rho_s \mathbf{u}^* \ddot{\mathbf{u}} \mathrm{d}V + \int_{\Omega_s} \boldsymbol{\varepsilon}^* \boldsymbol{\sigma} . \mathrm{d}V = \int_{\Omega_s} \rho_s \mathbf{u}^* \ddot{\mathbf{u}} \mathrm{d}V + \int_{\Omega_s} \boldsymbol{\varepsilon}^* \mathbf{D} \boldsymbol{\varepsilon} \mathrm{d}V = \int_{\Gamma_f} \mathbf{u}^* \bar{\mathbf{f}} \mathrm{d}S \qquad [4.23]$$

with:

$$\forall \mathbf{u}^* / \left\{ \mathbf{u}^* \big|_{\Gamma_u} = 0 \right\}.$$

4.5.3. *Finite element approximation*

As a first step, the domain Ω_s is subdivided into several finite elements (Ω_{se}); this meshing must be done with reference elements [MAK 15]. For each finite element, a number of points, known as nodes, is defined, to which degrees of freedom (nodal displacement) are assigned according to the adopted model (Table 4.1). The objective is then to define an approximation of solutions for each of these elements.

The sought-for approximations can be of polynomial type and must meet certain conditions: continuity, boundary conditions and completeness. This approximation is defined for the field of nodal displacement, which ensures the compatibility of displacements between adjacent elements and also the expression of the equilibrium conditions at the nodes. For each structural element, the nodal approximation of displacements can be written as follows:

$$\{\mathbf{u}\} = [\mathbf{N}_s]\{\mathbf{u}\}_e \qquad [4.24]$$

In this expression, $\{\mathbf{u}\}_e$ is the vector of nodal displacements of element e and $[\mathbf{N}_s]$ is the matrix of interpolation functions of a structural element. The domain Ω_s being discretized, the deformations/displacements relations are defined as a function of displacements at nodes by:

$$\{\varepsilon\}_e = [\mathbf{B}]_e \{\mathbf{u}\}_e \quad [4.25]$$

The matrix $[\mathbf{B}]_e$ is built using the interpolation functions. For the discretization of the integral function, the *Galerkin* method is most often used, which involves choosing for the fields of virtual displacement the same approximations chosen for the field of physical displacement (nodal displacement). This method has the advantage that it leads to systems of symmetric equations. Then, for each element $\boldsymbol{\Omega}_{se}$:

$$\int_{\Omega_{se}} \rho_s \mathbf{u}^* \ddot{\mathbf{u}} dV = \int_{\Omega_{se}} \rho_s \mathbf{u}^* [\mathbf{N}_s] \{\ddot{\mathbf{u}}\}_e dV = \int_{\Omega_{se}} \rho_s [\mathbf{N}_s]^t [\mathbf{N}_s] \{\ddot{\mathbf{u}}\}_e dV \quad [4.26]$$

$$\int_{\Omega_{se}} \boldsymbol{\varepsilon}^* \mathbf{D} \varepsilon dV = \int_{\Omega_{se}} \boldsymbol{\varepsilon}^* [\mathbf{D}][\mathbf{B}]_e \{\mathbf{u}\}_e dV = \int_{\Omega_{se}} [\mathbf{B}]_e^t [\mathbf{D}][\mathbf{B}]_e \{\mathbf{u}\}_e dV \quad [4.27]$$

The terms of equations [4.26] and [4.27] are complicated polynomials. Their analytical integration is not obvious. There are numerical integration methods that can be used to build elementary matrices by integration over the reference element, and the most common of them is the *Gauss* integration method [MAK 15].

$[\mathbf{M}]_e$ and $[\mathbf{K}]_e$ are respectively the elementary mass and rigidity matrices obtained after the numerical integration of [4.37] and [4.38]:

$$\int_{\Omega_{se}} \rho_s [\mathbf{N}_s]^t [\mathbf{N}_s] \{\ddot{\mathbf{u}}\}_e dV = [\mathbf{M}]_e \{\ddot{\mathbf{u}}\}_e \quad [4.28]$$

$$\int_{\Omega_{se}} [\mathbf{B}]_e^t [\mathbf{D}][\mathbf{B}]_e \{\mathbf{u}\}_e dV = [\mathbf{K}]_e \{\mathbf{u}\}_e \quad [4.29]$$

The next stage consists of assembling elementary matrices (of each element) to compose the global mass and rigidity matrices $[\mathbf{M}]$ and $[\mathbf{K}]$ (of the whole structure). This finally leads to the matrix writing [4.41] that represents the equation of motion of the structure:

$$[\mathbf{M}]\{\ddot{\mathbf{u}}\}+[\mathbf{K}]\{\mathbf{u}\}=\{\mathbf{f}\} \quad [4.30]$$

where $\{\mathbf{f}\}$ is the vector of the forces applied to nodes.

4.5.4. *Resolution in the frequency domain*

Let us consider the case of a sinusoidal excitation of angular frequency ω:

$$\{\mathbf{f}\}=\{\mathbf{f}(t)\}=\{\mathrm{f}_0\}\mathrm{e}^{i\omega t} \quad [4.31]$$

The nodal displacements have, by necessity, also a sinusoidal form:

$$\{\mathbf{u}\}=\{\mathbf{u}(t)\}=\{\mathbf{u}_0\}\mathrm{e}^{i\omega t} \quad [4.32]$$

In this case, the equation of motion becomes:

$$\left([\mathbf{K}]-\omega^2[\mathbf{M}]\right)\{\mathbf{u}\}=\{\mathbf{f}\} \quad [4.33]$$

4.5.4.1. *Calculation of eigenfrequencies (modal analysis)*

Modal analysis consists of assuming that the structure is not subjected to any excitation, and then solving the following equation with eigenvalues/eigenvectors:

$$\left([\mathbf{K}]-\omega_i^2[\mathbf{M}]\right)\{\boldsymbol{\varphi}_i\}=\{0\} \quad [4.34]$$

where:

- ω_i is the frequency of mode i;
- $\boldsymbol{\varphi}_i$ is the vector of nodal displacements of mode i.

It is important to note that there are as many modes as degrees of freedom.

4.5.4.2. *Calculation of the frequency response function*

The calculation of the FRF (Frequency Response Function) of a structure following a harmonic excitation goes through the calculation of the Transfer Function $[\mathbf{H}]$:

$$[\mathbf{H}] = \left([\mathbf{K}] - \omega^2 [\mathbf{M}]\right)^{-1} \quad [4.35]$$

4.6. Stochastic modal analysis of mechatronics components

Knowledge of mechanical properties is not perfect, being limited by uncertainties. These uncertainties related to geometry, properties of materials, fixing conditions or applied loading play an important role in the validation of numerical simulation models. The adopted approach aims to better evaluate the hazard of the response of numerical models in order to obtain a better match between the experimental measurements and the numerical results. The Monte Carlo method can be used to have n random draws, a numerical calculation being made for each draw in order to estimate the response to this game of values. The n calculations made for the n draws can be used to calculate estimators (mean, standard deviation, etc.) of the probabilistic response. The global sampling method by Latin hypercube is used in the Monte Carlo method [MAK 15]. Figure 4.13 shows the methodology adopted for the probabilistic estimation of a mechanical response (displacement, eigenfrequency, plastic deformation, etc.).

Sensitivity analysis aims to determine the influence of input variables on the response. The results obtained by the Monte Carlo method are used for the calculation of sensitivity indices. Sensitivity analysis relies on the calculation of linear correlation coefficients (or Pearson's correlation coefficients).

The estimation of moments (average and variance) of the FRF (respectively eigenfrequencies or dynamic response) of a mechanical system can be obtained by the Monte Carlo simulation [MAK 15]. Despite its heavy cost in terms of calculation time, this classical method is widely employed by specialized software (such as ANSYS) and serves as reference for approximate calculations. The FRF X is seen as a random variable, image of the basic random variables. The simulations consist of building a sample $(\mathrm{X}_1, \mathrm{X}_2, \ldots, \mathrm{X}_n)$ of the random variable X and process this sample with the usual statistical techniques. The n simulations are conducted independently, according to the law of distribution of basic random variables.

– The average of X is given by: $\mathrm{E}[\mathrm{X}] = \frac{1}{\mathrm{n}} \sum_{i=1}^{\mathrm{n}} \mathrm{X}_i$ [4.36]

– The variance of X is given by: $\mathrm{Var}[X] = \frac{1}{n-1}\sum_{i=1}^{n}[X_i - E(X)]^2$ [4.48]

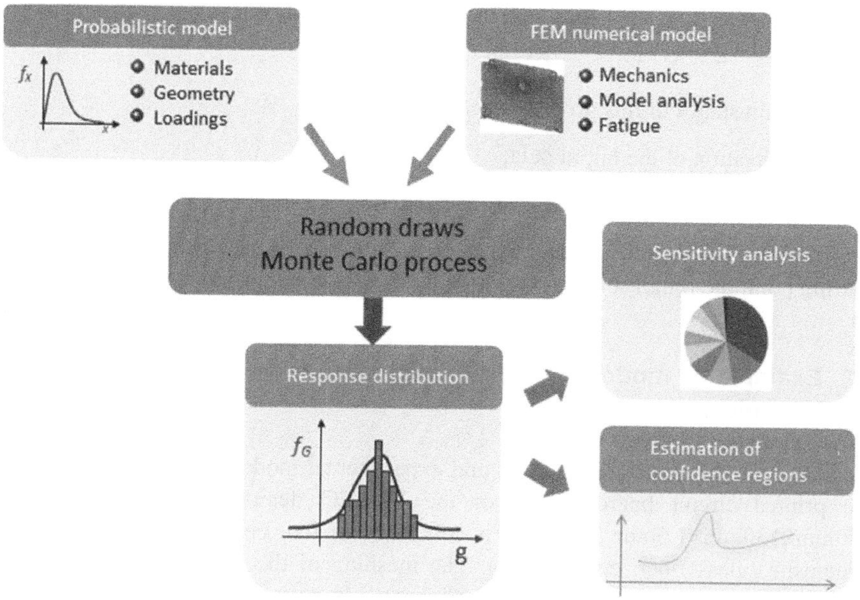

Figure 4.13. *Principle of stochastic analysis*

4.7. Numerical identification of the elastic parameters of electronic components

This study concerns the identification of elastic parameters (Young's modulus, Poisson's ratio and density) of the materials used in electronic components. Indeed, knowing these quantities is required in the numerical simulation of their vibratory behavior.

Parametric identification consists of minimizing the gap (objective function) between numerical results and experimental results, which corresponds to the following optimization problem:

$$\min_x F_{obj}(x) = \min_x \sum_{i=1}^{n_m} \left(\frac{f_{\exp,i} - f_{num,i}(x)}{f_{\exp,i}} \right)^2 \qquad [4.37]$$

where x is the vector of parameters to be identified, n_m is the number of modes, and $f_{exp,i}$ and $f_{num,i}$ are respectively the experimental and numerical frequencies. In order to solve this optimization problem, a hybrid approach is used, combining the genetic algorithm and the Levenberg–Marquardt algorithm based on a calculation of gradients [MAK 15]. This approach makes use of the advantages of each method in order to find a global optimal solution within a shorter calculation time.

The main stages of this process are:

– initialization of the initial data;

– search for a local optimum using the genetic algorithm;

– refining the search using the Levenberg–Marquardt algorithm and taking as starting point the rough solution provided by the genetic algorithm.

4.8. Example of modeling and simulation of the vibratory behavior of mechatronic components

This study concerns a numerical and experimental modal analysis conducted on the printed circuit board (PCB) of the VALEO demonstrator of dimensions 170mm×130mm×1.6mm. The PCB is composed of four copper layers and an FR4 composite (glass fiber, epoxy resin). The meshing of the numerical model of the studied electronic board is presented in Figure 4.14.

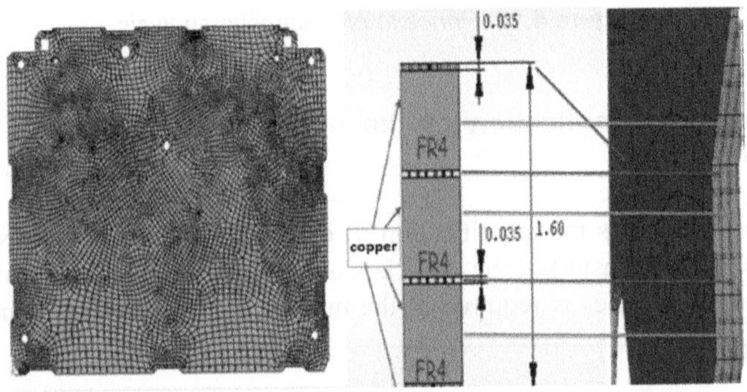

Figure 4.14. *Cross-section and Finite Element Model meshing of PCB*

The objective of the numerical and experimental study of the vibrations of the printed circuit is to examine the validity of the numerical modeling and verify the precision of the results obtained for modal deflections.

The works carried out are the following:

– numerical modal analysis;

– stochastic modal analysis;

– parametric identification and resetting with respect to experimental results;

– harmonic analysis using the identified parameters.

The numerical study is conducted using a generalist finite element calculation code (ANSYS code). The characteristics of the finite element model of the structure are the following:

– dimensions: 170x130x1.6 mm3;

– material properties: the deterministic simulation is made using the average values of the initial characteristics (before identification) of the PCB materials, these values being detailed in Table 4.2;

– boundary condition: fixed on the edge of five holes by fixing screws.

Description and unit	Variable	Average	Standard deviation	Distribution
Young's modulus FR4 (GPa)	YOUNG1	17	1.7	Normal
Young's modulus Cu (GPa)	YOUNG2	110	11	Normal
Shear modulus FR4 (GPa)	MO_CESA1	3	0.3	Normal
Shear modulus Cu (GPa)	MO_CEAS2	40	4	Normal
Poisson coefficient FR4 (GPa)	COF_POI1	0.14	0.014	Normal
Poisson's ratio Cu (GPa)	COF_POI2	0.35	0.035	Normal
Density FR4 (kg/m^3)	DENS1	1,900	190	Normal
Density Cu (kg/m^3)	DENS2	8,930	893	Normal

Table 4.2. *Description of the parameters of materials and their statistical distributions*

The modal analysis of the plate involves solving the following symmetric eigenvalues problem:

$$\left([K]-\omega^2[M]\right)\{u\}=0$$

The eigenvalues problem is symmetric, and the Cholesky algorithm [MAK 15] is used to extract the eigenmodes of the system. The corresponding modal deflections in the (o, z) plane are given in Figure 4.15.

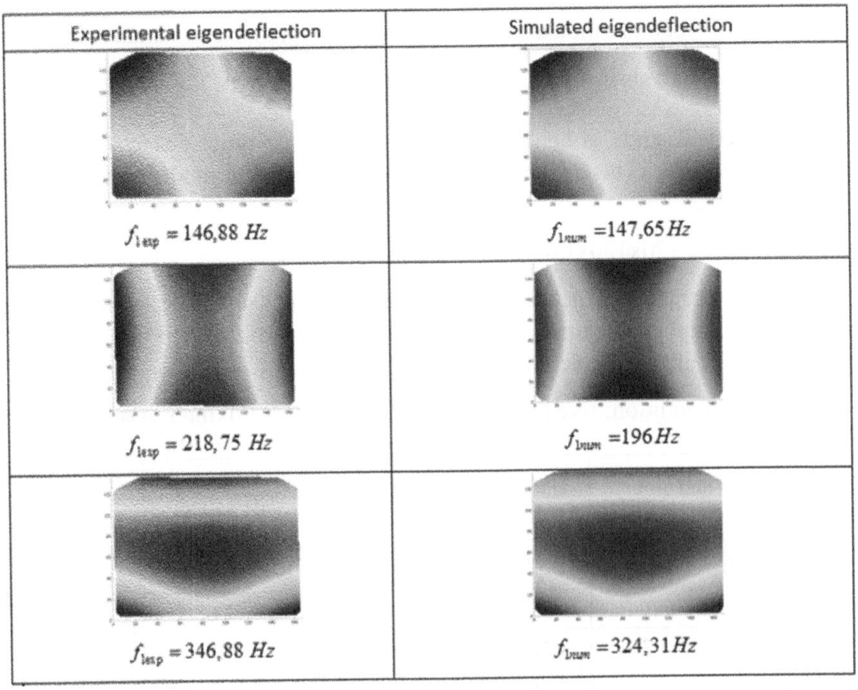

Figure 4.15. *Comparison of frequencies and experimental and numerical eigendeflections of PCB*

The results of the obtained simulations (eigenfrequencies and modal deflections) are compared to the experimental results obtained with the LDV (Laser Doppler Velocimetry) technique used by CEVAA (Centre d'Études Vibro-Acoustique pour l'Automobile; Center for acoustic and vibratory failure analysis) in Rouen. Figure 4.15 shows a good agreement between experimental and numerical results. This analysis allowed us to validate the numerical model. Moreover, the relative error between the first simulated and measured eigenfrequency is 3.7%, the state of the art accepts gaps up to 7% between numerical modeling and experimental measurements. This gap makes it possible to take uncertainties into consideration in a flat-rate manner. However, the objective of the stochastic modal analysis is to take into consideration uncertainties in the numerical modeling allowing a better concordance between experimental measurements and numerical results.

Imperfect knowledge on the parameters of the properties of the materials influences the quality of the resetting of the numerical model with respect to experimental results. In other terms, the uncertainties on the characteristics of materials and on the boundary conditions influence the frequencies and the eigendeflections of the structure. Moreover, these uncertainties also exist in the conditions in which the experimental devices are realized. In order to take these uncertainties into account, a stochastic modeling approach was previously proposed in order to estimate the probabilistic response of the variable of interest of the model (i.e. eigenfrequencies). A confidence region can be used to frame the calculated results with respect to the measured results. The stochastic analysis of eigenfrequencies is made with the Monte Carlo method. A total of 10,000 random draws of input data were made to calculate the first two statistical moments of the first 10 eigenfrequencies. However, an attempt can be made to precisely determine the numerical values of various parameters of materials in order to obtain a good concordance between the experimental results and those of the numerical simulation, this stage consisting of the resolution of the parametric identification problem by an optimization procedure. The sensitivity calculation process highlights the parameters with the highest influence on the response in order to reduce the number of parameters in the optimization process. A harmonic analysis is conducted using the identified parameters in order to simulate the FRF, allowing in particular the resetting to the experimentally measured FRF. Figure 4.16 shows the flow chart of the adopted methodology.

Table 4.3 presents the results of the stochastic modal analysis of PCB. These results show the dispersion of eigenvalues. The value of the first experimental eigenfrequency ranges within the region of confidence of 95% of the numerical modal analysis; this confidence region of [140.12 Hz; 154.07Hz] is calculated using the statistical moments obtained in Table 4.3 and the Bienaymé–Chebyshev inequality.

The linear correlation coefficients also known as Pearson's coefficients can be used for ranking the input parameters of the model depending on their importance. Figure 4.17 shows the sensitivities of the first eigenfrequency with respect to the input variables. These sensitivities allowed the consideration of the parameters influencing the simulation model in order to identify the best values that minimize the gap between the simulated and the experimental results. Moreover, the analysis of stochastic sensitivity was used to reduce the number of parameters to be identified, in order to accelerate and ensure the process of convergence of the optimization algorithm of the identification process. Indeed, the objective of this stage of study is to estimate the numerical values to be assigned to various parameters of the materials of the printed circuit, in order to obtain the best resetting of the experimental and simulated results. As expected, convergence is reached after a significant number of iterations, the optimal parameters of the printed circuit are presented in Table 4.4.

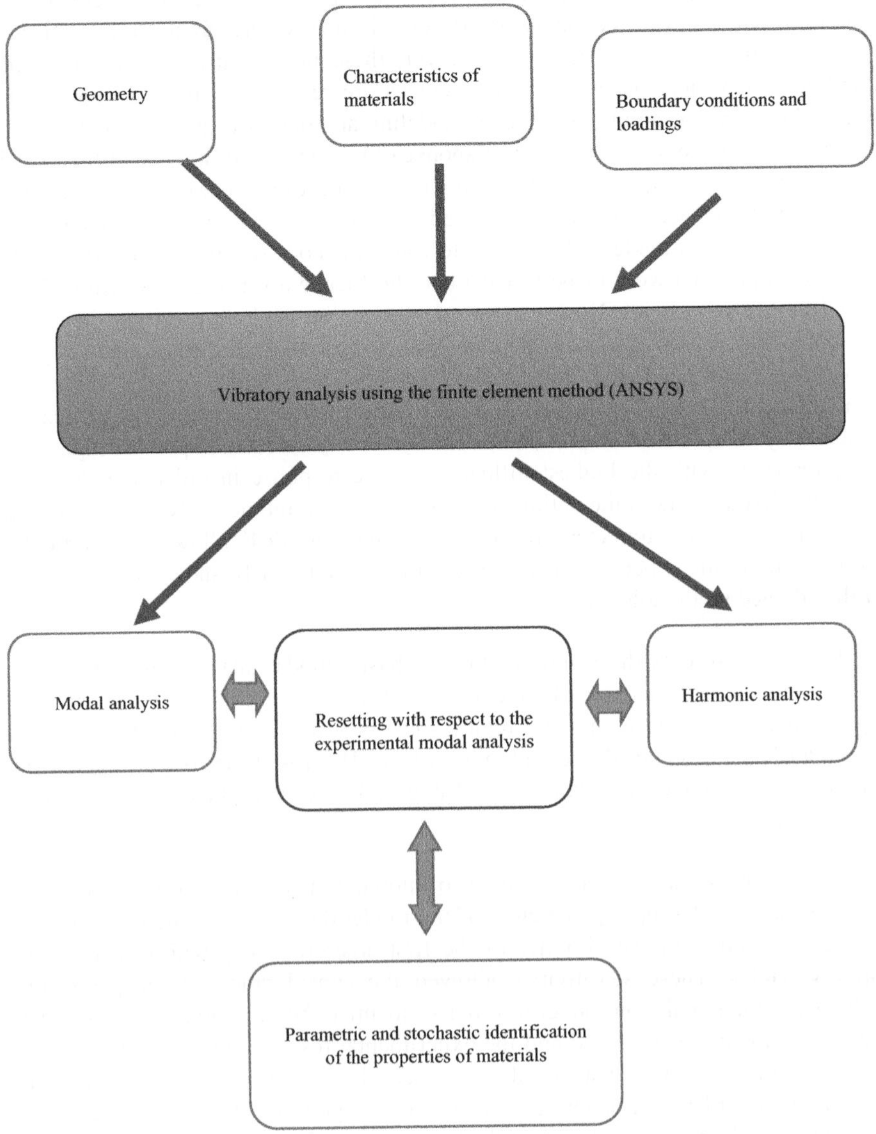

Figure 4.16. *Flow chart of the methodology adopted for the vibratory calculation*

	Experimental (CEVAA)	Numerical	
Mode	Frequency	Average	Standard deviation
1	146.9	147.1 (0.1%)	6.89
2	218.8	203.8 (6.0%)	9.34
3	346.9	331.1 (4.5%)	16.31
4	381.3	372.4 (2.3%)	16.81
5	453.1	435.6 (3.8%)	19.94
6	631.3	629.2 (0.3%)	29.49
7	756.3	749.1 (0.9%)	32.10
8	934.4	921 (1.4%)	38.29
9	1,069	1,052 (1.6%)	48.32
10	1,169	1,156 (1.1%)	51.12

Table 4.3. *Statistical moments of the first ten eigenfrequencies of PCB*

Parameters	Before identification	After identification
Young1 (GPa)	17	17.34
Young2 (GPa)	110	145.3
Dens1 (kg/m^3)	1,900	1,881
Dens2 (kg/m^3)	8,930	8,939

Table 4.4. *Identified parameters of VALEO PCB.*

Conducting a modal analysis using the newly identified parameters allowed a better resetting to measures. Table 4.5 shows the improvement of the resetting of simulated eigenfrequencies to measured values, where the maximal error is 3.5%. However, taking into account the shear modulus in the identification process can reduce this gap, since the frequencies due to the torsion mode are the least reset with respect to the experimental frequencies.

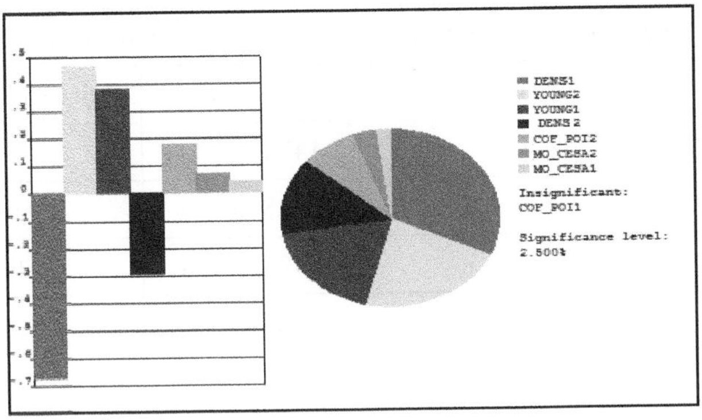

Figure 4.17. *Sensitivity of the first eigenfrequency of VALEO PCB*

Mode	Measured	Calculated
1	146.9	147.25 (0.2%)
2	218.8	210.46 (3.5%)
3	346.9	346.70 (0.0%)
4	381.3	378.31 (0.8%)
5	453.1	445.43 (1.7%)
6	631.3	629.24 (0.7%)
7	756.3	741.68 (1.9%)
8	934.4	938.82 (0.4%)
9	1,069	1,068.20 (0.0%)
10	1,169	1,169.78 (0.0%)

Table 4.5. *Results of the modal analysis after reset*

The objective of the harmonic analysis is to calculate the frequency response of the structure. Modal analysis can be used to calculate the eigenfrequencies and the eigendeflections, while harmonic analysis is used to obtain the displacements, velocities and accelerations of the structure as a function of the exciting frequency. The sinusoidal excitation is a nodal loading applied to the structure. The board is experimentally excited using a piezoelectric actuator stuck to the board [MAK 15]. At the modeling level, this artificial excitation is represented by a dynamic loading, modeled by forces focusing several nodes, in agreement with the experimental device (Figure 4.18). The dynamic force is given by the following relation:

$$F_{dyn} = \pm 4\pi^2 * m_{eff} * \frac{(\Delta_L)}{2} * f^2$$

where:

- F_{dyn} is the dynamic force in N;
- m_{eff} is the mass of the piezoelectric part in kg;
- Δ_L is the peak-to-peak displacement in m;
- f is the frequency in Hz.

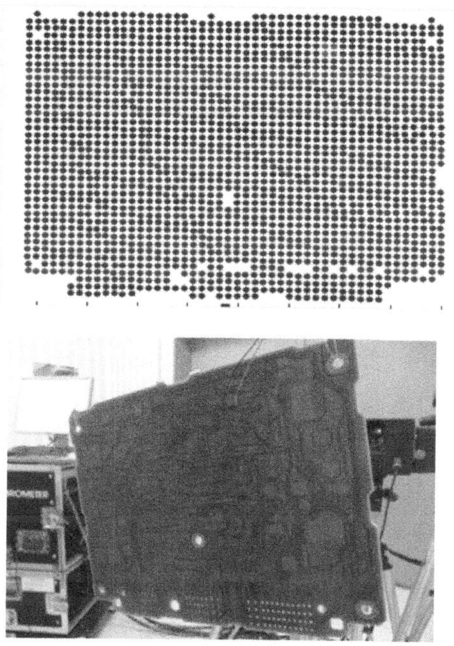

Figure 4.18. *Experimental meshing of the PCB and location of the exciter on the board*

4.9. Conclusion

Embedded electronic systems are used in various domains such as the automobile industry, astronautics, telecommunications, home appliances and the medical field. These embedded systems undergo many extreme stresses of a thermal nature (temperature variations), mechanical nature (collisions, vibration), etc. These stresses may lead to destructive failures in certain places. The numerical evaluation of vibratory and electro-thermal deformations becomes an important factor for a reliability study of electronic systems. This chapter presented the various deterministic and probabilistic numerical approaches for the study of the vibratory behavior and electro-thermo-mechanical behavior of electronic components.

Appendices

Appendix 1

Physical Properties of Common Fluids

Temperature	Density	Mass heat capacity	Thermal conductivity	Dynamic viscosity	Kinematic viscosity	Prandtl number
T °C	ρ $kg\ m^{-3}$	c $J\ kg^{-1}$	λ $W\ m^{-2}\ K^{-1}$	μ Pl or $Pa\ s^{-1}$	ν $m^2\ s^{-1}$ or stokes (stk)	Pr
0	1,002	4,218	0.552	17.9	1.31	13.06
20	1,001	4,182	0.597	10.1	1.43	7.02
40	995	4,178	0.628	6.55	1.51	4.34
60	985	4,184	0.651	4.71	1.55	3.02
80	974	4,196	0.668	3.55	1.64	2.22
100	960	4,216	0.68	2.82	1.68	1.74
120	945	4,250	0.685	2.33	1.71	1.45
140	928	4,283	0.684	1.99	1.72	1.24
160	910	4,342	0.68	1.73	1.73	1.1
180	889	4,417	0.675	1.54	1.72	1
200	867	4,505	0.665	1.39	1.71	0.94
220	842	4,610	0.652	1.26	1.68	0.89
240	816	4,756	0.635	1.17	1.64	0.88
260	786	4,949	0.611	1.08	1.58	0.87
280	753	5,208	0.58	1.02	1.48	0.91
300	714	5,728	0.54	0.96	1.32	1.02

Table A1.1. *Physical properties of water*

Temperature	Density	Mass heat capacity	Thermal conductivity	Dynamic viscosity	Kinematic viscosity	Prandtl number
T °C	ρ kg m^{-3}	c J kg^{-1}	λ W m^{-2} K^{-1}	μ Pl or Pa s^{-1}	ν m^2 s^{-1} or stokes (stk)	Pr
0	1.292	1,002	0.0242	1.72	1.86	0.72
20	1.204	1,006	0.0257	1.81	2.12	0.71
40	1.127	1,007	0.0272	1.9	2.4	0.7
60	1.059	1,008	0.0287	1.99	2.69	0.7
80	0.999	1,010	0.0302	2.09	3	0.7
100	0.946	1,012	0.0318	2.18	3.32	0.69
120	0.898	1,014	0.0333	2.27	3.66	0.69
140	0.854	1,016	0.0345	2.34	3.98	0.69
160	0.815	1,019	0.0359	2.42	4.32	0.69
180	0.779	1,022	0.0372	2.5	4.67	0.69
200	0.746	1,025	0.0386	2.57	5.05	0.68
220	0.700	1,028	0.0399	2.64	5.43	0.68
240	0.688	1,032	0.0412	2.72	5.8	0.68
260	0.662	1,036	0.0425	2.79	6.2	0.68
280	0.638	1,040	0.0437	2.86	6.59	0.68
300	0.616	1,045	0.0450	2.93	6.99	0.68

Table A1.2. *Physical properties of air*

Appendix 2

Physical Properties of Common Solids

	Density	Mass heat capacity	Thermal conductivity
	ρ $kg\ m^{-3}$	c $J\ kg^{-1}$	λ $W\ m^{-1}\ K^{-1}$
Carbon steel	7,833	465	54
Stainless steel 15% Cr. 10% Ni	7,864	460	20
Stainless steel 18% Cr, 8% Ni	7,816	460	16.3
Stainless steel 25% Cr, 20% Ni	7,864	460	13
Aluminum	2,707	896	204
Silver	10,525	234	407
Bronze 75% Cu, 25% Sn	8,800	377	188
Bronze 92% Cu, 8% AI	7,900	377	71
Carbon graphite	2,250	707	147
Chrome	2,118	7,160	449
Constantan 60% Cu, 40% Ni	8,922	410	22.7
Copper	8,954	383	386
Cupronickel 70% Cu, 30% Ni	8,900	377	29.3
Duralumin	2,787	3	164
Tin	7,304	226	64
Iron	7,870	452	73
Cast iron	7,849	460	59
Brass 70% Cu, 30% Zn	8,522	385	111
Magnesium	1,740	1,004	151
Gold	19,300	128	312
Platinum	21,400	140	69
Lead	11,373	130	35
Liquid sodium	930	1,381	84.5
Titanium	4,500	523	20.9
Tungsten	19,350	134	163
Zinc	7,144	384	112

Table A2.1. *Metallic materials: alloys*

	Density	Mass heat capacity	Thermal conductivity
	ρ $kg\,m^{-3}$	c $J\,kg^{-1}$	λ $W\,m^{-1}\,K^{-1}$
Asbestos	575	1,046	0.15
Asphalt	2,115	920	0.062
Rubber (natural)	1,150		0.28
Rubber (vulcanized)	1,100	2,010	0.13
Cardboard	86	2,030	0.048
Ice	920	2,040	1.88
Plexiglass	1,190	1,465	0.19
Porcelain	2,400	1,088	1.035
Polyethylene	929	1,830	0.46
PVC	1,459	930	0.21
Sand	1,515	800	0.2–1.0
Teflon	2,170	1,004	0.25
Wet earth	1,900	2,000	2
Dry earth	1,500	1,900	1
Glass	2,300	837	1.05
Pyrex glass	2,220	728	1.13

Table A2.2. *Miscellaneous materials*

	Density	Mass heat capacity	Thermal conductivity
	ρ $kg\,m^{-3}$	c $J\,kg^{-1}$	λ $W\,m^{-1}\,K^{-1}$
Slate	2,400	879	2.2
Basalt	2,850	881	1.6
Cavernous concrete	1,900	879	1.4
Solid concrete	2,300	878	1.75
Asphalt (cardboard)	1,050	1,305	0.23
Light hardwoods	525	3,143	0.15
Medium hardwoods	675	3,156	0.23
Very light hardwoods	375	3,147	0.12
Light softwoods	375	3,147	0.12
Medium–heavy softwoods	500	3,160	0.15
Very light softwoods	375	3,147	0.12
Clay brick	1,800	878	1.15
Hard limestone	2,450	882	2.4
Soft limestone	1,650	879	1
Tile	2,400	875	2.4

Okoume plywood	400	3,000	0.12
Pine plywood	500	3,000	0.15
Granite	2,600	881	3
Gravel (loose)	1,800	889	0.7
Sandstone	2,500	880	2.6
Lava	2,350	881	1.1
Marble	2,700	881	2.5
Gypsum	1,440	840	0.48
Shale	2,400	879	2.2

Table A2.3. *Materials used in construction*

	Density	**Mass heat capacity**	**Thermal conductivity**
	ρ $kg\,m^{-3}$	c $J\,kg^{-1}$	λ $W\,m^{-1}\,K^{-1}$
Balsa	140		0.054
Cotton	80	1,300	0.06
Rock wool (according to density)	20	880	0.047
	55	880	0.038
	135	880	0.041
Glass wool (according to density)	8	875	0.051
	10	880	0.045
	15	880	0.041
	40	880	0.035
Expanded cork	120	2,100	0.044
Carpet	200	1,300	0.06
Polyurethane foam (according to density)	32	1,300	0.03
	50	1,360	0.035
	85	1,300	0.045
PVC. rigid foam (according to density)	30	1,300	0.031
	40	1,300	0.041
Expanded polystyrene (according to density)	12	1,300	0.047
	14	1,300	0.043
	18	1,300	0.041

Table A2.4. *Insulation materials*

Appendix 3

Thermodynamic Properties of Water Vapor

For each temperature T of the vaporization stage, the table gives the saturating vapor tension p_{Sat}, the density of the liquid v_L, the density of the vapor v_V and the latent vaporization heat L.

T, °C	p_{Sat}, kPa	v_L, $kg.m^{-3}$	v_V, $kg.m^{-3}$	L, $kJ.kg^{-1}$
0.01	0.6113	0.001000	206.14	2501.3
5	0.8721	0.001000	147.12	2489.6
10	1.2276	0.001000	106.38	2477.7
15	1.7051	0.001001	77.93	2465.9
20	2.339	0.001002	57.79	2454.1
25	3.169	0.001003	43.36	2442.3
30	4.246	0.001004	32.89	2430.5
35	5.628	0.001006	25.22	2418.6
40	7.384	0.001008	19.52	2406.7
45	9.593	0.001010	15.26	2394.8
50	12.349	0.001012	12.03	2382.7
55	15.758	0.001015	9.568	2370.7
60	19.940	0.001017	7.671	2358.5
65	25.03	0.001020	6.197	2346.2

T, °C	p_{Sat}, MPa	v_L, $kg.m^{-3}$	v_V, $kg.m^{-3}$	L, J
70	31.19	0.001023	5.042	2333.8
75	38.58	0.001026	4.131	2321.4
80	47.39	0.001029	3.407	2308.8
85	57.83	0.001033	2.828	2296.0
90	70.14	0.001036	2.361	2283.2
95	84.55	0.001040	1.982	2270.2
100	0.10135	0.001044	1.6729	2257.0
105	0.12082	0.001048	1.4194	2243.7
110	0.14327	0.001052	1.2102	2230.2
115	0.16906	0.001056	1.0366	2216.5
120	0.19853	0.001060	0.8919	2202.6
125	0.2321	0.001065	0.7706	2188.5
130	0.2701	0.001070	0.6685	2174.2
135	0.3130	0.001075	0.5822	2159.6
140	0.3613	0.001080	0.5089	2144.7
145	0.4154	0.001085	0.4463	2129.6
150	0.4758	0.001091	0.3928	2114.3
155	0.5431	0.001096	0.3468	2098.6
160	0.6178	0.001102	0.3071	2082.6
165	0.7005	0.001108	0.2727	2066.2
170	0.7917	0.001114	0.2428	2049.5
175	0.8920	0.001121	0.2168	2032.4
180	1.0021	0.001127	0.19405	2015.0
185	1.1227	0.001134	0.17409	1997.1
190	1.2544	0.001141	0.15654	1978.8
195	1.3978	0.001149	0.14105	1960.0
200	1.5538	0.001157	0.12736	1940.7
205	1.7230	0.001164	0.11521	1921.0
210	1.9062	0.00117	0.10441	1900.7

T, °C	p_{Sat}, MPa	v_L, $kg.m^{-3}$	v_V, $kg.m^{-3}$	L, J
215	2.104	0.001181	0.09479	1879.9
220	2.318	0.001190	0.08619	1858.5
225	2.548	0.001199	0.07849	1836.5
230	2.795	0.001209	0.07158	1813.8
235	3.060	0.001219	0.06537	1790.5
240	3.344	0.001229	0.05976	1766.5
245	3.648	0.001240	0.05471	1741.7
250	3.973	0.001251	0.05013	1716.2
255	4.319	0.001263	0.04598	1689.8
260	4.688	0.001276	0.04221	1662.5
265	5.081	0.001289	0.03877	1634.4
270	5.499	0.001302	0.03564	1605.2
275	5.942	0.001317	0.03279	1574.9
280	6.412	0.001332	0.03017	1543.6
285	6.909	0.001348	0.02777	1511.0
290	7.436	0.001366	0.02557	1477.1
295	7.993	0.001384	0.02354	1441.8
300	8.581	0.001404	0.02167	1404.9
305	9.202	0.001425	0.019948	1366.4
310	9.856	0.001447	0.018350	1326.0
315	10.547	0.001472	0.016867	1283.5
320	11.274	0.001499	0.015488	1238.6
330	12.845	0.001561	0.012996	1140.6
340	14.586	0.001638	0.010797	1027.9
350	16.513	0.001740	0.008813	893.4
360	18.651	0.001893	0.006945	720.5
370	21.03	0.001213	0.004925	441.6
374.14	22.09	0.001155	0.003155	0

Appendix 4

Table of Functions: $erf(x)$. $erfc(x)$ and $ierfc(x)$

The three functions *erf*(*x*), *erfc*(*x*) and *ierfc*(*x*) are fundamental in the treatment of one-dimensional unsteady conduction problems. They are defined by:

$$erf(x) = \frac{2}{\pi}\int_0^x \exp(-\xi^2)\,d\xi$$

$$erfc(x) = 1 - \frac{2}{\pi}\int_0^x \exp(-\xi^2)\,d\xi$$

$$ierfc(x) = \frac{1}{\sqrt{\pi}}\exp(-x^2) - x\,erfc(x)$$

x	*erf*(*x*)	*erfc*(*x*)	*ierfc*(*x*)	*x*	*erf*(*x*)	*erfc*(*x*)	*ierfc*(*x*)
0	0	1	0.56418958	**1**	0.84270079	0.15729921	0.05025454
0.05	0.05637198	0.94362802	0.51559947	**1.1**	0.88020507	0.11979493	0.03646538
0.1	0.11246292	0.88753708	0.46982209	**1.2**	0.91031398	0.08968602	0.02604895
0.15	0.16799597	0.83200403	0.42683646	**1.3**	0.93400794	0.06599206	0.01831432
0.2	0.22270259	0.77729741	0.38660791	**1.4**	0.95228512	0.04771488	0.01267002
0.25	0.27632639	0.72367361	0.34908866	**1.5**	0.96610515	0.03389485	0.00862286
0.3	0.32862676	0.67137324	0.31421848	**1.6**	0.97634838	0.02365162	0.00577194
0.35	0.37938205	0.62061795	0.28192557	**1.7**	0.98379046	0.01620954	0.0037993

0.4	0.42839236	0.57160764	0.25212759	**1.8**	0.9890905	0.0109095	0.00245876
0.45	0.47548172	0.52451828	0.22473278	**1.9**	0.99279043	0.00720957	0.00156419
0.5	0.52049988	0.47950012	0.19964123	**2**	0.99532227	0.00467773	0.00097802
0.55	0.56332337	0.43667663	0.17674618	**2.1**	0.99702053	0.00297947	0.00060095
0.6	0.60385609	0.39614391	0.15593537	**2.2**	0.99813715	0.00186285	0.00036282
0.66	0.64937669	0.35062331	0.13354955	**2.3**	0.99885682	0.00114318	0.0002152
0.7	0.67780119	0.32219881	0.12009827	**2.4**	0.99931149	0.00068851	0.00012539
0.75	0.71115563	0.28884437	0.10483226	**2.5**	0.99959305	0.00040695	7.1762E-05
0.8	0.74210096	0.25789904	0.09117366	**2.6**	0.99976397	0.00023603	4.0336E-05
0.85	0.77066806	0.22933194	0.07900271	**2.7**	0.99986567	0.00013433	2.2264E-05
0.9	0.79690821	0.20309179	0.06820168	**2.8**	0.99992499	7.5013E-05	1.2067E-05
0.95	0.82089081	0.17910919	0.05865589	**2.9**	0.9999589	4.1098E-05	6.4216E-06
1	0.84270079	0.15729921	0.05025454	**3**	0.99997791	2.209E-05	3.355E-06

These functions are presented in Figure A4.1.

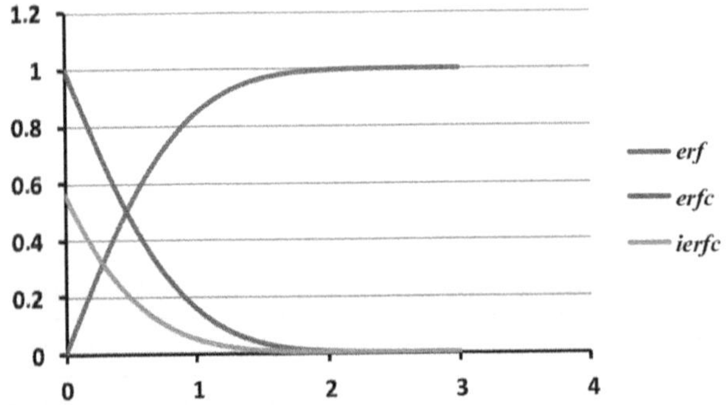

Figure A4.1. *Functions $erf\,\xi$, $erfc\,\xi$ and $ierfc\,\xi$. For a color version of this figure, see www.iste.co.uk/ledoux/heat4.zip*

References

[ABO 11a] ABO AL-KHEER A., EID M., AOUES Y. et al., "Theoretical analysis of the spatial variability in tillage forces for fatigue analysis of tillage machines", *Journal of Terramechanics*, vol. 48, no. 4, pp. 285–295, 2011.

[ABO 11b] ABO AL-KHEER A., KHARMANDA M.G., EL HAMI A. et al., "Estimating the variability of tillage forces on a chisel plough shank by modeling the variability of tillage system parameters", *Computers and Electronics in Agriculture*, vol. 78, no. 1, pp. 61–70, 2011.

[ABO 11c] ABO AL-KHEER A., EL-HAMI A., KHARMANDA M.G. et al., "Reliability-based design for soil tillage machines", *Journal of Terramechanics*, vol. 48, no. 1, pp. 57–64, 2011.

[ANT 05] ANTONOVA E.E., LOOMAN D.C., "Finite elements for thermoelectric device analysis in ANSYS", *ICT 2005 24th International Conference on Thermoelectrics*, pp. 215–218, 2005.

[ASS 17] ASSI I., MJALLAL I., FARHAT H. et al., "Using phase change material in heat sinks to cool electronics devices with intermittent usage", *IEEE 7th International Conference on Power and Energy Systems (ICPES)*, pp. 66–69, 2017.

[BAB 14] BABY R., BALAJI C., "Thermal performance of a MCP heat sink under different heat loads: An experimental study", *International Journal of Thermal Sciences*, vol. 79, pp. 240–249, 2014.

[BAL 11] BALAJI C., MUNGARA P., SHARMA P., "Optimization of size and shape of composite heat sinks with phase change materials", *Heat and Mass Transfer*, vol. 47, no. 15, pp. 597–608, 2011.

[BEN 61] BENNETT J.A.R., COLLIER J.G., PRATT H.C.R., THORNTON J.D., "Heat transfer to two phase gas-liquid systems. Pt 1: Steam water mixtures in the liquid dispersed region in an annulus", *Trans Inst Chem Engrs*, vol. 39, no. 113, or AERE-R-3519, 1961.

[BEN 13] BEN ABDESSALEM A., PAGNACCO E., EL-HAMI A., "Increasing the stability of T-shape tube hydroforming process under stochastic framework", *International Journal of Advanced Manufacturing Technology*, vol. 69, nos 5–8, pp. 1343–1357, 2013.

[BEN 15] BEN ABDESSALEM A., EL HAMI A., "A probabilistic approach for optimising hydroformed structures using local surrogate models to control failures", *International Journal of Mechanical Sciences*, vols 96–97, pp. 143–162, 2015.

[BER 64] BERGLES A.E., ROSENHOW W.M., "The determination of forced convection surface boiling heat transfer", *Trans, ASME Journ of Heat Transfer*, vol. 86, no. 3, pp. 365–372, 1964.

[BER 81] BERGLES A.E., COLLIER J.G., DELHAYE J.M. et al., *Two-Phase Flow and Heat Transfer in the Power and Process Industries*, McGraw Hill, New York, 1981.

[BER 11] BERGMAN T.L., INCROPERA F.P., DE WITT, D.P. et al., *Fundamentals of Heat and Mass Transfer*, John Wiley & Sons, New York, 2011.

[BOU 96] BOUYEKHF R., EL MOUDNI A., EL HAMI A. et al., "Reduced order modelling of two-time-scale discrete non-linear systems", *Journal of the Franklin Institute*, vol. 333, no. 4, pp. 499–512, 1996.

[BOU 01] BOUYEKHF R., EL HAMI A., EL MOUDNI A., "Optimal control of a particular class of singularly perturbed nonlinear discrete-time systems", *IEEE Transactions on Automatic Control*, vol. 46, no. 7, pp. 1097–1101, 2001.

[BRU 65] BRUN E.A., *Théorie de la couche limite*, Gauthier Villars, Paris, 1965.

[BYR 66] BYRON BIRD R., STEWART W.E., LIGHTFOOT, E.N., *Transport Phenomena*, John Wiley & Sons, New York, 1966.

[CEN 15] CENGEL Y., GHAJAR A., *Heat and Mass Transfer: Fundamentals and Applications*, 5th ed., McGraw-Hill Education, New York, 2015.

[CHE 08] CHEROUAT A., RADI B., EL HAMI A., "The frictional contact of the composite fabric's shaping", *Acta Mechanica*, vol. 199, nos 1–4, pp. 29–41, 2008.

[CHI 11] CHINTAKRINDA K., WEINSTEIN R., FLEISCHER A., "A direct comparison of three different material enhancement methods on the transient thermal response of paraffin phase change material exposed to high heat fluxes", *International Journal of Thermal Sciences*, vol. 50, no. 19, pp. 1639–1647, 2011.

[CIA 88] CIARLET P., *Introduction à l'analyse numérique matricielle et à l'optimisation*, Masson, Paris, 1988.

[COS 10] COSKUN A., ATIENZA D., ROSING T. et al., "Energy-efficient variable-flow liquid cooling in 3D stacked architectures", *Proceedings of the Conference on Design, Automation and Test in Europe, European Design and Automation Association*, pp. 111–116, 2010.

[DAM 17] DAMMAK K., EL HAMI A., KOUBAA S. et al., "Reliability based design optimization of coupled acoustic-structure system using generalized polynomial chaos", *International Journal of Mechanical Sciences*, vol. 134, pp. 75–84, 2017.

[DAM 19] DAMMAK, K., KOUBAA, S., EL HAMI, A. et al., "Numerical modelling of vibro-acoustic problem in presence of uncertainty: Application to a vehicle cabin", *Applied Acoustics*, vol. 144, pp. 113–123, 2019.

[DAV 66] DAVIS E.J, ANDERSON G.H., "The incipience of nucleate boiling in forced convection flow" *A.I.C.H.E. Journal*, vol. 12, no. 4, pp. 774–780, 1966.

[DEB 21] DEBICH B., Modélisation, simulation, et fiabilité des matériaux changement de phase pour l'automobile et l'aéronautique, Thesis, INSA Rouen Normandie, 2021.

[DEN 56] DENGLER C.E., ADDOMS J.N., "Heat transfer mechanism for vaporisation of water in a vertical tube", *Chem Engng Prog Symp*, Series 52, no. 18, pp. 95–103, 1956.

[DHA 84] DHATT Y., TOUZOT G., *Une présentation de la méthode des éléments finis*, Maloine, Paris, 1984.

[ECK 72] ECKERT, E.R.G., DRAKE, R.M., *Analysis of Heat and Mass Transfer*, McGraw-Hill, New York, 1972.

[EDM 68] EDMOND A., BRUN A., MARTINOT L. et al., *Mécanique des fluides Tomes I, II et III*, Dunod, 1968.

[ELH 93] EL HAMI A., LALLEMENT G., MINOTTI P. et al., "Methods that combine finite group theory with component mode synthesis in the analysis of repetitive structures", *Computers and Structures*, vol. 48, no. 6, pp. 975–982, 1993.

[ELH 96] EL HAMI A., RADI B., "Some decomposition methods in the analysis of repetitive structures", *Computers and Structures*, vol. 58, no. 5, pp. 973–980, 1996.

[ELH 09] EL HAMI A., RADI B., CHEROUAT A., "Treatment of the composite fabric's shaping using a Lagrangian formulation", *Mathematical and Computer Modelling*, vol. 49, nos 7–8, pp. 1337–1349, 2009.

[ELH 11] EL HAMI A., RADI B., "Comparison study of different reliability-based design optimization approaches", *Advanced Materials Research*, vol. 274, pp. 113–121, 2011.

[ELH 13a] EL HAMI A., RADI B., *Incertitude et optimisation et fiabilité des structures*, Hermès-Lavoisier, Paris, 2013.

[ELH 13b] EL HAMI A., RADI B., *Uncertainty and Optimization in Structural Mechanics*, ISTE, London, and John Wiley & Sons, New York, 2013.

[FOK 10] FOK S., SHEN W., TAN F., "Cooling of portable hand-held electronic devices using phase changematerials in finned heat sinks", *International Journal of Thermal Sciences*, vol. 49, no. 11, pp. 109–117, 2010.

[FOS 98] FOSSETT A., MAGUIRE M., KUDIRKA A. et al., "Avionics passive cooling with microencapsulated phase change materials", *Journal of Electronic Packaging*, vol. 120, no. 13, pp. 238–242, 1998.

[FRA 95] FRANC J.P., AVELLAN, F., BELAHADJI, B. et al., *La cavitation. Mécanismes physiques et aspects industriels*, Collection Grenoble Sciences, EDP Sciences, Grenoble, 1995.

[GUE 56] GUERRIERI S.A., TALTY R.D., "A study of heat transfer to organic liquid in single tube natural circulation vertical tube boilers", *Chem Engng Prog Symp, Series Heat Transfer*, Louisville, vol. 52, no. 18, pp. 67–77, 1956.

[GUE 15a] GUERINE A., EL HAMI A., WALHA L. et al., "Approach for the dynamic analysis of one stage gear system with uncertain parameters", *Mechanism and Machine Theory*, vol. 92, pp. 113–126, 2015.

[GUE 15b] GUERINE A., EL HAMI A., FAKHFAKH T. et al., "A polynomial chaos method to the analysis of the dynamic behavior of spur gear system", *Structural Engineering and Mechanics*, vol. 53, no. 4, pp. 819–831, 2015.

[GUE 16a] GUERINE A., EL HAMI A., WALHA L. et al., "A polynomial chaos method for the analysis of the dynamic behavior of uncertain gear friction system", *European Journal of Mechanics, A/Solids*, vol. 59, pp. 76–84, 2016.

[GUE 16b] GUERINE A., HAMI A.E., WALHA L. et al., "Dynamic response of a Spur gear system with uncertain friction coefficient", *Advances in Engineering Software*, vol. 120, pp. 45–54, 2016.

[GUE 17] GUERINE A., EL HAMI A., WALHA L. et al., "Dynamic response of wind turbine gear system with uncertain-but-bounded parameters using interval analysis method", *Renewable Energy*, vol. 113, pp. 679–687, 2017.

[HAR 56] HARTNETT E.B., ECKERT E.R.G., BIRKEBAK R., SAMPSON R.L., "Simplified procedure for the calculation of heat transfer to surfaces with non-uniform temperature", *WADC Technical Report*, vol. 56, no. 373, December 1956.

[HAS 16] HASAN A., HEJASE H., ABDELBAQI S. et al., "Comparative effectiveness of different phase change materials to improve cooling performance of heat sinks for electronic devices", *Applied Sciences*, vol. 6, no. 19, p. 226, 2016.

[HEL 03] HELTON J.C., DAVIS F.J., "Latin hypercube sampling and the propagation of uncertainty in analyses of complex systems", *Reliability Engineering & System Safety*, vol. 81, no. 1, pp. 23–69, 2003.

[HEN 19] HENK T., LUMLEY J., *A First Course in Turbulence*, MIT Press, Cambridge, 2019.

[HER 55] HERMANN S., *Boundary-Layer Theory*, Springer Verlag, Berlin, 1955.

[HEW 69] HEWITT G.F., ROBERTS D.N., Studies of two-phase flow patterns by simultaneous X-ray and flash photography, AERE-M2159, 1969.

[HOD 02] HODES M., WEINSTEIN R.D., PENCE S.J. et al., "Transient thermal management of a handset using phase change material (MCP)", *Journal of Electronic Packaging*, vol. 124, no. 14, pp. 419–426, 2002.

[HUA 16] HUANG, C., RADI, B., EL HAMI, A., "Uncertainty analysis of deep drawing using surrogate model based probabilistic method", *Manufacturing Technology*, vol. 86, nos 9–12, pp. 3229–3240, 2016.

[JAN 17] JANNOUN M., AOUES Y., PAGNACCO E. et al., "Probabilistic fatigue damage estimation of embedded electronic solder joints under random vibration", *Microelectronics Reliability*, vol. 78, pp. 249–257, 2017.

[JEN 51] JEN W.H., LOTTES P.A., Analysis of heat transfer burnout, pressure drop, and density data for high pressure water, ANL-4627, 1951.

[KAN 07] KANDASAMY R., WANG X., MUJUMDAR A., "Application of phase change materials in thermal management of electronics", *Applied Thermal Engineering*, vol. 27, nos 117–118, pp. 2822–2832, 2007.

[KAN 08] KANDASAMY R., WANG X., MUJUMDAR A., "Transient cooling of electronics using phase change material", *Applied Thermal Engineering*, vol. 28, nos 8–9, pp. 1047–1057, 2008.

[KRI 04] KRISHNAN S., GARIMELLA S.V., "Analysis of a phase change energy storage system for pulsed power dissipation", *IEEE Transactions on Components and Packaging Technologies*, vol. 27, no. 11, pp. 191–199, 2004.

[LAN 84] LANDAU L.D., LIFSHITZ E.M., *Electrodynamics of Continuous Media*, 2nd ed., vol. 8, Butterworth-Heinemann, Oxford, 1984.

[LED 75] LEDOUX M., Simulation thermique d'écoulements avec réaction de surface. Etude de la couche limite de diffusion en régime laminaire et turbulent, PhD thesis, Université de Rouen, 1975.

[LED 17a] LEDOUX M., EL HAMI A., *Fluid Mechanics Analytical Methods*, ISTE Ltd, London, and John Wiley & Sons, New York, 2017.

[LED 17b] LEDOUX M., EL HAMI A., *Fluid Mechanics, Compressible Flow, Propulsion and Digital Approaches in Fluid Dynamics*, ISTE Ltd, London, and John Wiley & Sons, New York, 2017.

[LED 21a] LEDOUX M., EL HAMI A., *Heat Transfer 1, Conduction*, ISTE, Ltd, London and John Wiley & Sons, New York, 2021.

[LED 21b] LEDOUX M., EL HAMI A., *Heat Transfer 2, Radiative Transfer*, ISTE, Ltd, London and John Wiley & Sons, New York, 2021.

[LEO 00] LEONI, N., AMON C., "Bayesian surrogates for integrating numerical, analytical, and experimental data: Application to inverse heat transfer in wearable computers", *IEEE Transactions on Components and Packaging Technologies*, vol. 23, pp. 23–32, 2000.

[LIG 50] LIGHTHILL M.J., "Contribution to the theory of heat transfer through a laminar boundary layer", *Proceedings of the Royal Society A*, pp. 202–359, 1950.

[MAB 17] MABROUK, I.B., EL HAMI, A., WALHA, L. et al., "Dynamic vibrations in wind energy systems: Application to vertical axis wind turbine", *Mechanical Systems and Signal Processing*, vol. 85, pp. 396–414, 2017.

[MAK 16] MAKHLOUFI A., AOUES Y., EL HAMI A., "Reliability based design optimization of wire bonding in power microelectronic devices", *Microsystem Technologies*, vol. 22, no. 12, pp. 2737–2748, 2016.

[MJA 18] MJALLAL I., FARHAT H., HAMMOUD M. et al., "Improving the cooling efficiency of heat sinks through the use of different types of phase change materials", *Technologies*, vol. 1, no. 15, p. 6, 2018.

[MOR 02] MORO T., EL HAMI A., EL MOUDNI A., "Reliability analysis of a mechanical contact between deformable solids", *Probabilistic Engineering Mechanics*, vol. 17, no. 3, pp. 227–232, 2002.

[NIS 03] NISTOR I., PANTALÉ O., CAPERAA S. et al., "Identification of a dynamic viscoplastic flow law using a combined Levenberg-Marquardt and Monte-Carlo algorithm", *VII International Conference on Computational Plasticity COMPLAS*, Barcelona, 2003.

[NUK 66] NUKIYAMA S., "The maximum and minimum value of the heat Q transmitted from metal to boiling water under atmospheric pressure", *International of Heat and Mass Transfer*, vol. 9, pp. 1419–1433, 1966.

[NYE 57] NYE J.F., *Physical Properties of Crystals: Their Representation by Tensors and Matrices*, Clarendon Press, Oxford, 1957.

[PAL 98] PAL D., JOSHI Y., "Thermal management of an avionics module using solid-liquid phase-change materials", *Journal of Thermophysics and Heat Transfer*, vol. 12, no. 12, pp. 256–262, 1998.

[PY 01] PY X., OLIVES R., MAURAN S., "Paraffin/porous-graphite-matrix composite as a high and constant power thermal storage material", *International Journal of Heat and Mass Transfer*, vol. 44, no. 114, pp. 2727–2737, 2001.

[RAD 07] RADI B., EL HAMI A., "Reliability analysis of the metal forming process", *Mathematical and Computer Modelling*, vol. 45, nos 3–4, pp. 431–439, 2007.

[ROS 52] ROSENHOW W.M., "A method of correlating heat-transfer data for surface boiling of liquids", *Trans. ASME*, vol. 74, pp. 969–976, 1952.

[RUB 51] RUBESIN M.W., The effect of arbitrary surface temperature variation along a flat plate on the convective heat transfer in an incompressible turbulent boundary layer, NANA TN 2345, 1951.

[SAR 11] SARSRI, D., AZRAR, L., JEBBOURI, A. et al., "Component mode synthesis and polynomial chaos expansions for stochastic frequency functions of large linear FE models", *Computers and Structures*, vol. 89, nos 3–4, pp. 346–356, 2011.

[SCH 62] SCHROCK V.E., GROSSMAN L.M., "Forced convection boiling in tubes", *Nuclear Science and Engineering*, vol. 12, pp. 474–480, 1962.

[SEB 51] SEBAN R.A., Experimental investigation of convective heat transfer on a pipe to air from a flat plate with stepwise discontinuous surface temperature, Thesis, University of California, Berkeley, 1951.

[SHI 72] SHINOZUKA M., "Monte Carlo solution of structural dynamics", *International Journal Computer Structural*, vol. 2, pp. 855–874, 1972.

[SLA 71] SLAUGHTERBACK D.C., YBARRONDO L.J., OBENCHAIN C.F., "Flow filmboiling heat transfer correlations-parametrice study with data comparison", *ASME-AICHE Heat Transfer Conference,* Atlanta, 1971.

[SUS 11] SUSMAN G., DEHOUCHE Z., CHEECHERN T. et al., "Tests of prototype MCP 'sails' for office cooling", *Applied Thermal Engineering*, vol. 31, no. 15, pp. 717–726, 2011.

[TIA 11] TIAN Y., ZHAO C.Y., "A numerical investigation of heat transfer in phase change materials (MCPs) embedded in porous metals", *Energy*, vol. 36, no. 19, pp. 5539–5546, 2011.

[WAR 98] WARREN M., ROSENHOW J.P., HARTNETT Y.I. et al., *Hand Book of Heat Transfer*, McGraw Hill, New York, 1998.

[YAN 09] YANG Y.T., PENG H.S., "Investigation of planted pin fins for heat transfer enhancement in plate fin heat sink", *Microelectronics Reliability*, vol. 49, no. 12, pp. 163–169, 2009.

[ZHE 04] ZHENG, N., WIRTZ R.A., "A hybrid thermal energy storage device, part 1: Design methodology", *Journal of Electronic Packaging*, vol. 126, no. 11, pp. 1–7, 2004.

[ZHO 11] ZHOU D., ZHAO C.Y., "Experimental investigations on heat transfer in phase change materials (MCPs) embedded in porous materials", *Applied Thermal Engineering*, vol. 31, no. 15, pp. 970–977, 2011.

Index

A, B

C, D

E, F

G, H

K, M

N, O

P, R

S, T, V

Other titles from

iSTE

in

Mechanical Engineering and Solid Mechanics

2022

BAYLE Franck
Product Maturity 1: Theoretical Principles and Industrial Applications (Reliability of Multiphysical Systems Set – Volume 12)
Product Maturity 2: Principles and Illustrations (Reliability of Multiphysical Systems Set – Volume 13)

LANNOY André
Reliability of Nuclear Power Plants: Methods, Data and Applications (Reliability of Multiphysical Systems Set – Volume 14)

LEDOUX Michel, EL HAMI Abdelkhalak
Heat Transfer 3: Convection, Fundamentals and Monophasic Flows (Mathematical and Mechanical Engineering Set – Volume 11)

2021

CHALLAMEL Noël, KAPLUNOV Julius, TAKEWAKI Izuru
Modern Trends in Structural and Solid Mechanics 1: Static and Stability
Modern Trends in Structural and Solid Mechanics 2: Vibrations
Modern Trends in Structural and Solid Mechanics 3: Non-deterministic Mechanics

DAHOO Pierre Richard, POUGNET Philippe, EL HAMI Abdelkhalak
Applications and Metrology at Nanometer Scale 1: Smart Materials, Electromagnetic Waves and Uncertainties
(Reliability of Multiphysical Systems Set – Volume 9)
Applications and Metrology at Nanometer Scale 2: Measurement Systems, Quantum Engineering and RBDO Method
(Reliability of Multiphysical Systems Set – Volume 10)

LEDOUX Michel, EL HAMI Abdelkhalak
Heat Transfer 1: Conduction
(Mathematical and Mechanical Engineering Set – Volume 9)
Heat Transfer 2: Radiative Transfer
(Mathematical and Mechanical Engineering Set – Volume 10)

PLANCHETTE Guy
Cindynics, The Science of Danger: A Wake-up Call
(Reliability of Multiphysical Systems Set – Volume 11)

2020

SALENÇON Jean
Elastoplastic Modeling

2019

BAYLE Franck
Reliability of Maintained Systems Subjected to Wear Failure Mechanisms: Theory and Applications
(Reliability of Multiphysical Systems Set – Volume 8)

BEN KAHLA Rabeb, BARKAOUI Abdelwahed, MERZOUKI Tarek
Finite Element Method and Medical Imaging Techniques in Bone Biomechanics
(Mathematical and Mechanical Engineering Set – Volume 8)

IONESCU Ioan R., QUEYREAU Sylvain, PICU Catalin R., SALMAN Oguz Umut
Mechanics and Physics of Solids at Micro- and Nano-Scales

LE VAN Anh, BOUZIDI Rabah
Lagrangian Mechanics: An Advanced Analytical Approach

MICHELITSCH Thomas, PÉREZ RIASCOS Alejandro, COLLET Bernard, NOWAKOWSKI Andrzej, NICOLLEAU Franck
Fractional Dynamics on Networks and Lattices

SALENÇON Jean
Viscoelastic Modeling for Structural Analysis

VÉNIZÉLOS Georges, EL HAMI Abdelkhalak
Movement Equations 5: Dynamics of a Set of Solids
(Non-deformable Solid Mechanics Set – Volume 5)

2018

BOREL Michel, VÉNIZÉLOS Georges
Movement Equations 4: Equilibriums and Small Movements
(Non-deformable Solid Mechanics Set – Volume 4)

FROSSARD Etienne
Granular Geomaterials Dissipative Mechanics: Theory and Applications in Civil Engineering

RADI Bouchaib, EL HAMI Abdelkhalak
Advanced Numerical Methods with Matlab® 1: Function Approximation and System Resolution
(Mathematical and Mechanical Engineering SET – Volume 6)
Advanced Numerical Methods with Matlab® 2: Resolution of Nonlinear, Differential and Partial Differential Equations
(Mathematical and Mechanical Engineering SET – Volume 7)

SALENÇON Jean
Virtual Work Approach to Mechanical Modeling

2017

BOREL Michel, VÉNIZÉLOS Georges
Movement Equations 2: Mathematical and Methodological Supplements
(Non-deformable Solid Mechanics Set – Volume 2)

Movement Equations 3: Dynamics and Fundamental Principle
(Non-deformable Solid Mechanics Set – Volume 3)

BOUVET Christophe
Mechanics of Aeronautical Solids, Materials and Structures
Mechanics of Aeronautical Composite Materials

BRANCHERIE Delphine, FEISSEL Pierre, BOUVIER Salima, IBRAHIMBEGOVIC Adnan
From Microstructure Investigations to Multiscale Modeling: Bridging the Gap

CHEBEL-MORELLO Brigitte, NICOD Jean-Marc, VARNIER Christophe
From Prognostics and Health Systems Management to Predictive Maintenance 2: Knowledge, Traceability and Decision
(Reliability of Multiphysical Systems Set – Volume 7)

EL HAMI Abdelkhalak, RADI Bouchaib
Dynamics of Large Structures and Inverse Problems
(Mathematical and Mechanical Engineering Set – Volume 5)
Fluid-Structure Interactions and Uncertainties: Ansys and Fluent Tools
(Reliability of Multiphysical Systems Set – Volume 6)

KHARMANDA Ghias, EL HAMI Abdelkhalak
Biomechanics: Optimization, Uncertainties and Reliability
(Reliability of Multiphysical Systems Set – Volume 5)

LEDOUX Michel, EL HAMI Abdelkhalak
Compressible Flow Propulsion and Digital Approaches in Fluid Mechanics
(Mathematical and Mechanical Engineering Set – Volume 4)
Fluid Mechanics: Analytical Methods
(Mathematical and Mechanical Engineering Set – Volume 3)

MORI Yvon
Mechanical Vibrations: Applications to Equipment

2016

BOREL Michel, VÉNIZÉLOS Georges
Movement Equations 1: Location, Kinematics and Kinetics
(Non-deformable Solid Mechanics Set – Volume 1)

BOYARD Nicolas
Heat Transfer in Polymer Composite Materials

CARDON Alain, ITMI Mhamed
New Autonomous Systems
(Reliability of Multiphysical Systems Set – Volume 1)

DAHOO Pierre Richard, POUGNET Philippe, EL HAMI Abdelkhalak
Nanometer-scale Defect Detection Using Polarized Light
(Reliability of Multiphysical Systems Set – Volume 2)

DE SAXCÉ Géry, VALLÉE Claude
Galilean Mechanics and Thermodynamics of Continua

DORMIEUX Luc, KONDO Djimédo
Micromechanics of Fracture and Damage
(Micromechanics Set – Volume 1)

EL HAMI Abdelkhalak, RADI Bouchaib
Stochastic Dynamics of Structures
(Mathematical and Mechanical Engineering Set – Volume 2)

GOURIVEAU Rafael, MEDJAHER Kamal, ZERHOUNI Noureddine
From Prognostics and Health Systems Management to Predictive Maintenance 1: Monitoring and Prognostics
(Reliability of Multiphysical Systems Set – Volume 4)

KHARMANDA Ghias, EL HAMI Abdelkhalak
Reliability in Biomechanics
(Reliability of Multiphysical Systems Set –Volume 3)

MOLIMARD Jérôme
Experimental Mechanics of Solids and Structures

RADI Bouchaib, EL HAMI Abdelkhalak
Material Forming Processes: Simulation, Drawing, Hydroforming and Additive Manufacturing
(Mathematical and Mechanical Engineering Set – Volume 1)

2015

KARLIČIĆ Danilo, MURMU Tony, ADHIKARI Sondipon, MCCARTHY Michael
Non-local Structural Mechanics

SAB Karam, LEBÉE Arthur
Homogenization of Heterogeneous Thin and Thick Plates

2014

ATANACKOVIC M. Teodor, PILIPOVIC Stevan, STANKOVIC Bogoljub, ZORICA Dusan
Fractional Calculus with Applications in Mechanics: Vibrations and Diffusion Processes
Fractional Calculus with Applications in Mechanics: Wave Propagation, Impact and Variational Principles

CIBLAC Thierry, MOREL Jean-Claude
Sustainable Masonry: Stability and Behavior of Structures

ILANKO Sinniah, MONTERRUBIO Luis E., MOCHIDA Yusuke
The Rayleigh–Ritz Method for Structural Analysis

LALANNE Christian
Mechanical Vibration and Shock Analysis – 5-volume series – 3rd edition
Sinusoidal Vibration – Volume 1
Mechanical Shock – Volume 2
Random Vibration – Volume 3
Fatigue Damage – Volume 4
Specification Development – Volume 5

LEMAIRE Maurice
Uncertainty and Mechanics

2013

ADHIKARI Sondipon
Structural Dynamic Analysis with Generalized Damping Models: Analysis

ADHIKARI Sondipon
Structural Dynamic Analysis with Generalized Damping Models: Identification

BAILLY Patrice
Materials and Structures under Shock and Impact

BASTIEN Jérôme, BERNARDIN Frédéric, LAMARQUE Claude-Henri
Non-smooth Deterministic or Stochastic Discrete Dynamical Systems: Applications to Models with Friction or Impact

EL HAMI Abdelkhalak, RADI Bouchaib
Uncertainty and Optimization in Structural Mechanics

KIRILLOV Oleg N., PELINOVSKY Dmitry E.
Nonlinear Physical Systems: Spectral Analysis, Stability and Bifurcations

LUONGO Angelo, ZULLI Daniele
Mathematical Models of Beams and Cables

SALENÇON Jean
Yield Design

2012

DAVIM J. Paulo
Mechanical Engineering Education

DUPEUX Michel, BRACCINI Muriel
Mechanics of Solid Interfaces

ELISHAKOFF Isaac *et al.*
Carbon Nanotubes and Nanosensors: Vibration, Buckling and Ballistic Impact

GRÉDIAC Michel, HILD François
Full-Field Measurements and Identification in Solid Mechanics

GROUS Ammar
Fracture Mechanics – 3-volume series
Analysis of Reliability and Quality Control – Volume 1
Applied Reliability – Volume 2
Applied Quality Control – Volume 3

RECHO Naman
Fracture Mechanics and Crack Growth

2011

KRYSINSKI Tomasz, MALBURET François
Mechanical Instability

SOUSTELLE Michel
An Introduction to Chemical Kinetics

2010

BREITKOPF Piotr, FILOMENO COELHO Rajan
Multidisciplinary Design Optimization in Computational Mechanics

DAVIM J. Paulo
Biotribolgy

PAULTRE Patrick
Dynamics of Structures

SOUSTELLE Michel
Handbook of Heterogenous Kinetics

2009

BERLIOZ Alain, TROMPETTE Philippe
Solid Mechanics using the Finite Element Method

LEMAIRE Maurice
Structural Reliability

2007

GIRARD Alain, ROY Nicolas
Structural Dynamics in Industry

GUINEBRETIÈRE René
X-ray Diffraction by Polycrystalline Materials

KRYSINSKI Tomasz, MALBURET François
Mechanical Vibrations

KUNDU Tribikram
Advanced Ultrasonic Methods for Material and Structure Inspection

SIH George C. *et al.*
Particle and Continuum Aspects of Mesomechanics

Printed and bound by CPI Group (UK) Ltd, Croydon, CR0 4YY
16/09/2026
14963077-0001